# RESIDUE REVIEWS

RESIDUES OF PESTICIDES AND OTHER
FOREIGN CHEMICALS IN FOODS AND FEEDS

## RÜCKSTANDS-BERICHTE

RÜCKSTÄNDE VON PESTICIDEN UND ANDEREN
FREMDSTOFFEN IN NAHRUNGS- UND FUTTERMITTELN

EDITED BY

## FRANCIS A. GUNTHER

RIVERSIDE

VOLUME 5

SPRINGER-VERLAG
BERLIN · GÖTTINGEN · HEIDELBERG
1964

SPECIAL VOLUME – SYMPOSIUM ON

# INSTRUMENTATION
# FOR THE DETECTION AND DETERMINATION
# OF PESTICIDES AND THEIR RESIDUES
# IN FOODS

## LOS ANGELES MEETINGS
### OF THE
### AMERICAN CHEMICAL SOCIETY

### APRIL 1963

SPRINGER-VERLAG
BERLIN · GÖTTINGEN · HEIDELBERG
1964

ISBN 978-1-4615-8385-1          ISBN 978-1-4615-8383-7 (eBook)
DOI 10.1007/978-1-4615-8383-7

© by Springer-Verlag OHG
Berlin · Göttingen · Heidelberg 1964
Softcover reprint of the hardcover 1st edition 1964
Library of Congress Catalog Card Number 62—18595

# Preface

The object of "Residue Reviews" is to provide concise, critical reviews of timely advances, philosophy, and significant areas of accomplished or needed endeavor in the total field of residues of pesticide and certain other chemicals in foods, feeds, and in transformed food products. During the 144th National Meetings of the *American Chemical Society* in Los Angeles, California the *Pesticides Subdivision* of the A. C. S. *Division of Agricultural and Food Chemistry* on April 1, 1963 sponsored a symposium "Instrumentation for the Detection and Determination of Pesticides and Their Residues in Foods". With special permission from R. N. HADER and his associates in the *American Chemical Society*, that symposium is reproduced in this volume of "Residue Reviews" because of its timeliness and significance to pesticide residue analysts everywhere.

Without exception the symposium authors accepted my invitation to utilize "Residue Reviews" as their publication medium, and their cooperation in the extra chore of providing manuscripts conforming to the style requirements of "Residue Reviews" is gratefully acknowledged.

Presiding over the two sessions of the symposium were CHARLES L. DUNN and MILTON S. SCHECHTER, whose introductory remarks at that time comprise the foreword to this volume; their enthusiastic assistance both in organizing the symposium and in achieving this final product is warmly appreciated.

F. A. G.

Department of Entomology
University of California

Riverside, California
June 27, 1963

# Foreword

This symposium is part of the continuing effort of the *Pesticides Sub-division* of the *American Chemical Society's* Division of Agricultural and Food Chemistry to serve the various specialized chemical interests of its members. These interests reflect the overlapping of many chemical skills and fields of knowledge directed toward the use of chemicals to elicit biological responses that will aid man in his effort to control his environment or to help in his adaptation to it.

Francis A. Gunther, who organized this symposium, recognized more than a decade ago the profound effects that the widespread use of large variety of pesticide chemicals would have, that the ramifications would be many, and that some of these could be predicted, while others would be unexpected. The accuracy of this foresight is borne out by recent activities in the United States of bodies such as the *National Research Council*, the *United States Senate*, and the President's *Science Advisory Committee*, all of whom have undertaken to review the important rôle of pesticides in our agriculture and in our fight against pest-borne disease, as well as to assess problems attending the safe use of these products.

While the term "residue chemistry" is not an entirely satisfactory term for embracing the various types of investigations arising from the use of pesticides, it is sufficiently descriptive of an area of interest to warrant its use, provided the many connotations associated with it are kept in mind. Interest in the application of instrumental methods to the investigation of pesticide residue problems has been on the increase as reflected in past meetings of the *American Chemical Society* as well as in the excellent attendance at this Los Angeles meeting.

The essence of this symposium is the desire of scientists to share their knowledge and to increase their proficiency by gaining from the experience of others. The detection and determination of pesticides and their residues is such a complex matter that those who must choose methods for a particular study can rarely, if ever, be expert in all of the areas of instrumentation that might be brought to bear on their problems. The value of advanced instrumentation to analytical chemists depends on the degree to which certain improvements are attained. Improvements such as those in precision, accuracy, sensitivity, speed, specificity or resolving power, in the acquisition and display of desired qualitative or quantitative data, and in automation in handling samples are some of the major goals of newer instrumentation. Expressions of fact or even opinions based on experiences of competent investigators are of inestimable help to the person who is seeking to choose among new methods and instruments or to improve the application of those already at his disposal. Certainly, the avoidance of "booby traps" that may await the unwary investigator is reason enough to seek the counsel and observations of the most experienced investigators. The high expense of much of the instrumentation sold today makes errors in the selection of equipment for residue studies very costly, indeed. Above all, the constant striving of the residue chemist himself for excellence in his work leads to active participation in meetings such as this one.

The present publication of the entire group of papers making up this symposium is another manifestation of the excellent cooperation amongst the various groups involved in pesticide chemistry. Special acknowledgement is due RODNEY N. HADER, editor of the *Journal of Agricultural and Food Chemistry*, and to others in the *American Chemical Society* who so promptly and decisively acted upon the request that the present papers be released for publication outside the *Society* journals. The authors have similarly been helpful in readying their manuscripts for publication.

CHARLES L. DUNN
(Chairman, *Pesticides Subdivision*)
Hercules Powder Company
Wilmington, Delaware

MILTON S. SCHECHTER
(Chairman-Elect, *Pesticides Subdivision*)
Pesticide Chemicals Research Branch
U. S. D. A. Agricultural Research Center
Beltsville, Maryland

# Table of Contents

Introduction to symposium

# Special features in the analysis of pesticide residues: Residue analysis and food control

By

H. Frehse [*]

With 6 figures

It is indeed for me a great honor to be called upon to present the introductory paper at this symposium. I regret that I cannot appear personally. In my view, the purpose of this introduction is not to lecture on new techniques of residue analysis, which will be treated anyway in special papers. Furthermore, it would be tantamount to nothing more than a platitude if I were to point out to this group that the problems and techniques of residue analysis cannot be properly dealt with in thirty minutes. Comprehensive reports have recently been published which give a better presentation of this subject: in English by GUNTHER (1962), in German by ourselves (FREHSE and NIESSEN 1963), and in French by LEMOAN (1962).

The aim of my contribution is to broach upon a number of thoughts concerning the perspectives which appear to the residue analyst when he is faced with the task of ensuring that the requirements of food laws have been observed: in brief, to carry out market control on foods when complete spray history information is not available. Considering that such tasks must be completed within a short time so that food quality and factory operation will be unaffected, these thoughts will also be of general benefit to the residue analyst. After all, it is still a shortcoming of most specific residue methods that they take up too much time.

In Europe the treatment of pesticides residue questions in food legislation has thus far been a very slow procedure. The situation in 1961 was that Ireland, Yugoslavia, Luxembourg, Poland, Portugal, and Turkey had still issued no regulations (BERAN 1961). Tolerance lists are at present planned in Holland and in Western Germany (BERAN 1961). Furthermore Bulgaria, Czechoslovakia, Eastern Germany, Hungary, Poland, and Russia have formed a joint working committee, and have proposed that a uniform list of tolerances be accepted in those countries (*Tagungsberichte* 1962) [1]. Some of them have issued their own lists of tolerance recommendations,

[*] Biologisches Institut, Farbenfabriken Bayer A. G., Leverkusen-Bayerwerk, W. Germany.
[1] Not in effect yet.

although of no great length, while in Hungary regulations already exist on pre-harvest intervals (*Tagungsberichte* 1962).

Pre-harvest intervals are rightly given preference in most countries because tolerances are transcendent for farmers. Regulations so far issued urgently require coordination (cf. DURRENMATT 1962). The pre-harvest intervals laid down in the various countries often differ by as much as 150 percent: for Metasystox 42 days in Belgium and Holland, 21 days in Western Germany and in Great Britain; for aldrin and dieldrin 42 days in Belgium, 21 days in Norway; for DDT 35 days in Austria, 14 days in Great Britain (BERAN 1961; see also UNTERSTENHÖFER and FREHSE 1963). Such a situation can no longer be accounted for by climatic differences, different eating habits, or diverging crop-protection practices, but can only be attributed to shortcomings in administration. The basic principle underlying the preparation of lists of tolerances in Western Germany is that the tolerances should not be higher than the residue values that are found on crops under our biological, climatic, and cultivation conditions, and provided the pesticide has been properly applied (LEIB 1962 and 1963). These are the so-called "maximum tolerances". The analysis of transformation products of active ingredients, also mentioned in the German Food Law, is to be enforced at a later date on the grounds of special studies carried out by official bodies (LEIB 1962 and 1963). In Western Germany, just as in the rest of Europe, there are hardly any state departments, university institutes, or food inspection offices which have already specialized in residue problems. Much of the fundamental data originates from the manufacturers of the pesticides. There is still no uniform definition of the term "unpermitted residue". On the other hand, no European country has such an extensive control organization as exists in the United States. In our country, it is almost impossible for official organizations to know which pesticides have been applied by farmers on the different crops, in a certain growing area. Nevertheless, all civilized countries, whether they have lists of tolerances or not, will be faced in the near future by the same problem, namely of being able to prevent misuse of pesticides, and to detect, and ultimately prevent, undesired residues in foods.

This control of foods will, therefore, constitute a grave problem for the residue analyst. The preparation of analytical methods for newly developed compounds is no longer enough. Rather, the analyst must attempt to find methods of testing which will enable him to state which pesticides are present in the foods, in the form of unpermitted residues. The great need at the moment is for rapid, simplified screening, and sensitive qualitative procedures to identify the residues and to give at least semi-quantitative ideas of their amounts (SCHECHTER and WESTLAKE 1962; cf. ANONYMOUS 1961). It is obvious that the development of a residue analytical method and the study of the transformation of pesticides on and in the plant are inseparable. After all, to analyze residues means that one has to know what kind they are. Allowance for this doubling of purpose should also be made in the general methods which the residue analyst should attempt to devise in anticipating market controls.

There are, at present and in principle, three methodical possibilities for dealing with the problem of general methods:

1. *Chemical or physical group analyses,* by means of which a certain class of substances can be determined. These methods can hardly be applied without first cleaning up the plant extracts. A condition for their application is that the cleanup procedure can also be uniformly employed for the whole class of substances.

2. *Biological test methods,* to show the presence or absence of toxicologically significant residues.

3. *Chromatographic methods,* which can permit separation and simultaneous identification of any residues present. At the moment this group exclusively consists of gas-chromatographic methods for which new possibilities have been opened, at least for halogenated compounds, by the introduction of the well-known microcoulometric technique of COULSON and for these and some others by the well-known electron-capture detector of LOVELOCK and LIPSKY. The detection of organophosphates by these methods is still in the very early stages. These methods and other similar techniques will be lectured on by competent experts, so that I need not enter into details. I shall briefly refer to other chromatographic procedures.

The first attempt to make a *chemical group analysis* of residues of organophosphate insecticides was carried out in 1961 by LAWS and WEBLEY (1961). I should like to recall briefly the principle of their method as shown in Fig. 1 [2]. The first extraction is carried out by macerating the sliced vegetable with dichloromethane. After filtration, washing and evaporation of the solvent, partition is effected between light petroleum (40—60° C.) and a 15 percent solution of methanol in water. At this stage, the insecticides are distributed into solutions *A* and *B.* Solution *A* in light petroleum is chromatographed on a column of graded alumina (Brockman grade *V*). The water-soluble portion (*B*) is examined after elution with chloroform from a carbon column. The phosphorus content is determined in both groups of eluates. The authors maintain that the final eluates contain the insecticides substantially free from organic phosphorus of plant origin. This statement is based on the experience which these authors accumulated mainly with cabbage extracts. Application of the method in this form, however, is limited. Beet leaves which, because of their high phosphorus content provide a real test for the suitability of such a cleanup procedure, yielded plant blank values up to 6.6 p.p.m. in our experiments. It is especially in the

---

[2] Please note in the left-hand column the compound Bayer 29493 (common name: fenthion), the active ingredient of the insecticides "Lebaycid", "Entex", and "Tiguvon", on which we shall concentrate our attention at times for demonstration purposes:

$$(MeO)_2 \overset{S}{P}-O-\left\langle \quad \right\rangle \begin{array}{c} -SCH_3 \\ CH_3 \end{array}$$

detection of metabolites that the method may fail, however, as demonstrated in Fig. 2 by a case encountered in our own work (Niessen et al. 1962 a).

Sliced vegetable material

Macerate with dichloromethane
(All common insecticides extracted)

Partition between light petroleum and 15 percent methanol-water

*Petroleum-soluble* organophosphorus
insecticides (solution A)

Chromatography on alumina; light
petroleum as eluting agent

Chlorthion *           Wash column suc-
Disyston *            cessively with light
Diazinon             petroleum and a
Fenchlorphos          15 percent solution of
Parathion-methyl       diethyl ether in light
Phenkapton            petroleum
Phorate (Thimet *)
S 1752 = Bayer 29493    Gusathion *
                     Malathion

*Water-soluble* organophosphorus
insecticides (solution B)

Chromatography on carbon; chloroform as eluting agent

Demeton-methyl and metabolites
Dimefox
Morphothion
Phorate oxygen-analogue sulphone
Phosdrin *
Phosphamidon
Rogor *
Trichlorphon

* Trade name.

Fig. 1. Extraction and separation of residual insecticides from vegetables into water-soluble and petroleum-soluble groups, according to Laws and Webley (1961)

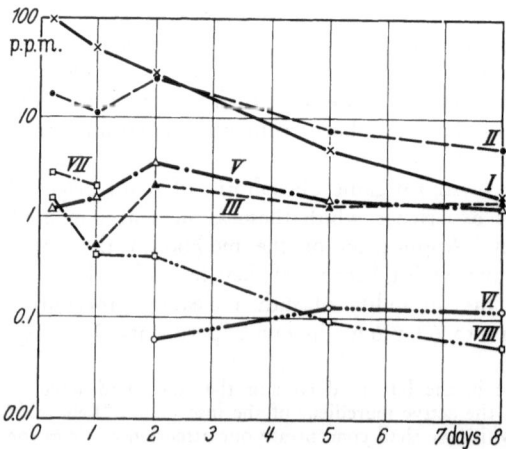

Fig. 2. Composition of chloroform extracts after application of P³²-labelled Lebaycid (*I*) on bean plants: days after application vs. p.p.m. (Niessen *et al.* 1962 a). See text for other identifications

In the plant, the active ingredient Bayer 29493 forms sulfoxides and sulfones as metabolites, both from the thionophosphate form *(II, III)* and the phosphate form *(V, VI);* also, the sulfoxide of the methyl-*S*-isomer form, which can be prepared by thermal rearrangement, could be detected *(VIII)* (NIESSEN *et al.* 1962 a). The maximum yield that can possibly be obtained by the method of LAWS and WEBLEY (1961) on the grounds of solubility ratios amount (FREHSE 1961) to:

33 percent for thiono-sulfoxide
80 percent for thiono-sulfone
16 percent for phosphate-sulfoxide
2 percent for phosphate-sulfone

In order to determine this active ingredient and these metabolites with an adequate yield of 75 to 80 percent at blank values of 0.2 to 0.3 p.p.m., we had to design a method which consists of six cleanup stages (FREHSE *et al.* 1962 b)! Of course, all "general methods" are only methods of approximation. With this proviso, the method of LAWS and WEBLEY (1961) still is important for semi-quantitative group determinations of organo-phosphate pesticide residues.

The value of any method which uses a microdetermination of *phosphorus* for its decisive measurement is governed by the reproducible amount of phosphorus of plant origin which survives the cleanup procedure. Plant material contains, among others, the following approximate amounts of phosphorus (calculations based on a 90 percent water content) (FREHSE and NIESSEN 1963):

potatoes . . . . . . . 175 p.p.m.
clover . . . . . . . . 240 p.p.m.
cabbage leaves . . . . . 300 p.p.m.
grass . . . . . . . . . 460 p.p.m.
beet leaves (greenhouse) . . 900 p.p.m.
hay . . . . . . . . . 2,400 to 3,000 p.p.m.

This can mean a 10,000-fold excess of phosphorus for residues of less than a part per million.

It is interesting to compare proportions of this phosphorus that are retained by chromatographic adsorbents. We introduced chloroform extracts of tomato leaves into columns containing different adsorbents, each filling 10 cm. high and 1.5 cm. in diameter, using aliquots representing 25 and 50 g. of plant tissue. The results are given in Table I (FREHSE 1963). The data in the last line represent a special case. They do not originate from a chromato-graphic cleanup, but were obtained after precipitating plant substances with a solution of phosphoric acid and ammonium chloride, then extracting the aqueous phase with chloroform (FREHSE and NIESSEN 1963).

Such a comparison may bring to light very considerable differences, as proved by the phosphorus-values in the last column. Such a series, which of course can be increased, can be made the basis of a limited group analysis when the active ingredients to be analyzed are compared with the object of establishing which of the adsorbents best allows each substance to pass through, and provided a suitable selection and combination of cor-responding separating columns can be made.

In this connection I would also remind you that the diazinon method of SUTER *et al.* (1955) (hydrolysis with hydrobromic acid solution, deter-mination of the resultant hydrogen sulfide as methylene blue) can be extended to other monothio- or dithiophosphate residues, e.g., dimethoate or Guthion. By this method, too, it is also possible to determine residues of a whole group of organophosphates without the necessity to carry out an

Table I. *Relative amounts of natural phosphorus compounds, of a raw chloroform extract from tomato leaves, retained by different chromatographic adsorbents* (FREHSE, 1963)

| Fraction analyzed | Eluting agent (100 ml. each) | Phosphorus content | |
|---|---|---|---|
| | | p.p.m. | % |
| Tomato leaves, unextracted . . . . . . . | — | 350 | — |
| Raw chloroform extract from tomato leaves[a] | — | 40 | = 100 |
| Eluate from column filled with: | | | |
| *Activated carbon* „for gas chromatography" (Merck), particle size 0.5 to 0.75 mm . . | Chloroform | — | 50 |
| *Aluminum oxide, impregnated* with paraffin/ vaseline[b] . . . . . . . . . . . . . | Acetonitrile/ water = 1 : 2 | — | 9.4 |
| *Activated carbon,* 14 to 20 mesh BBS (Messrs. Sutcliffe, Speakman of Leigh, Lancashire[c] | Chloroform | — | 7.3 |
| *Aluminum oxide, impregnated* (see above) . . . | Acetonitrile/ water = 2 : 1 | — | 2.0 |
| *Aluminum oxide,* neutral (Merck), Brockmann grade *I* . . . . . . . . . . . . . . | Chloroform | — | 0.4 |
| *Florisil,* 60/100 mesh . . . . . . . . . . | Chloroform | — | 0.3 |
| *Aluminum oxide,* neutral (Merck), Brockmann grade *V*. . . . . . . . . . . . . . | Chloroform | — | < 0.2 |
| *Carbon, pulverized,* Nuchar C 190-N (Fisher) . | Chloroform | — | < 0.2 |
| Chloroform extract from purified raw extract (as above). Purification by precipitation with phosphoric acid/ammonium chloride solution[d].. . . . . . . . . . . . . | — | — | < 0.2 |

[a] Chloroform-soluble phosphorus (compounds) in the tomato leaves.
[b] ERWIN et al. (1955).
[c] HANCOCK and LAWS (1955), TIETZ and FREHSE (1960).
[d] FREHSE and NIESSEN (1963).

all too comprehensive cleanup procedure. This method, however, is limited in that (1) any transformation products present in the form of phosphoric acid derivatives are not detected, and (2) it is only semi-quantitative for residues of unknown history, as illustrated with dithiophosphates where both $S$-atoms may react. This method should be preferred as a fast method of analysis in cases where the total residue contains only a slight percentage of phosphate metabolites, as with our active ingredient Bayer 29493 (MAIER-BODE 1962).

The possibilities for a general method are more favorable where residues of *carbamates* are involved. We have developed a cleanup procedure by means of which monomethyl carbamate residues can be jointly determined quantitatively by evaluating the $N$-$H$ stretching frequency at 2.88 $\mu$ ($\approx 3,500$ cm.$^{-1}$). By using ordinate expansion, we are able to attain a lower limit of detection of about 0.2 p.p.m., and the method is specific for monomethyl carbamate residues. We shall give detailed information on this method at the International Pesticides Congress in July, 1963 in London (see NIESSEN and FREHSE 1963). Jointly determined carbamates can also be separated by paper-chromatography and thin-layer chromatography, and individually analyzed with phenol reagents.

Generally speaking, the methods of *paper chromatography* or *thin-layer chromatography* are essential aids for identifying unknown pesticides from plant extracts. For example, we used two paper-chromatographic systems in clarifying the complex plant metabolism of the active ingredient of Bayer 29493, as already mentioned (cf. p. 3) (NIESSEN *et al.*

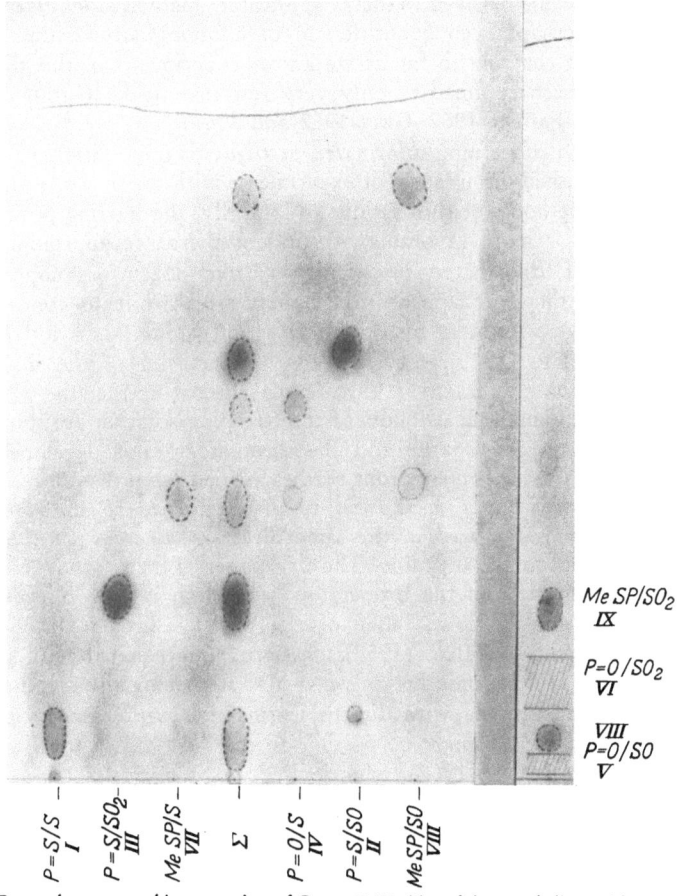

Fig 3. Paper chromatographic separation of Bayer 29493 (*I*) and its metabolites with two paper-and-solvent systems. For explanation of other numbers see text; *IX* = methyl-*S*-isomer sulfone

1962 b). On acetylated paper (acetylation grade 12 percent), most of the metabolites are well separated in the solvent system acetone/acetonitrile/water (1 : 1 : 3) as shown in the left-hand section of Fig. 3. The substances

are marked at the bottom edge by their essential groups. The compounds
V, VI, and VIII, positioned close to the front in this system, can be separated in a second system on paper impregnated with propylene-glycol (right-
hand section of Fig. 3). Unlike the other substances, the sulfoxide (V) and
the sulfone (VI) of the phosphate form cannot be rendered visible by
spraying with palladium chloride, or even be distinguished by means of
differential staining: they had to be located and identified by systematic
ashing of the paper.

These systems can also be used for other aromatic phosphoric esters.
But as interesting as they may be for the purpose of identifying, from a
mixture, individual active ingredients which form only one or two transformation products or none at all, prospects are nevertheless poor for
developing on this basis a widely applicable method for determining
natures, and perhaps even quantities, of organophosphate residues in the
sense of market control. So far as we know, experiments in this direction
have thus far been confined to only very few insecticides (Eichenberger
and Gay 1960, Faderl 1962, Getz 1962 and 1963).

The situation concerning chlorinated pesticides is quite different. Results
have been published on numbers of experiments with the object of developing screening methods for such residues. Naturally, the starting position for
such endeavors is more favorable. At present there are approximately a
dozen available chlorinated insecticides to five dozen organophosphate
insecticides. It therefore appears to be most worthwhile to continue the
efforts which were initiated by Mitchell (1958), McKinley and Mahon
(1959), and Mills (1959). I would also remind you of the interesting
studies by Major and Barry (1961). These authors extract the plants by
the Mills (1959) method. Aliquots of the stripping solution are spotted on
$8 \times 8$ inch ($20 \times 20$ cm.) paper and the chromatogram is developed with
acetonitrile. When the solvent front reaches a line drawn one inch (2.5 cm.)
from the top, the paper is allowed to dry, returned to the tank, and
chromatography is repeated in the same direction whereby the pesticides
are concentrated on the top line. The dry paper is then turned 180° and
standards are spotted on the former top line which is now considered to
be one inch from the lower edge of the paper. Chromatography then
proceeds according to Mills (1959), on paper impregnated with mineral
oil, 75 percent acetone or other solvents in water being the mobile phase,
or on paper impregnated with dimethylformamide with trimethylpentane
as the mobile phase. It might be possible to combine this method with the
experiences of Evans (1962). He devised a practical system for the
separation of eleven chlorinated insecticides or their metabolites by determining the experimental conditions leading to reproducible $R_F$ values and
area measurements, using a special chromogenic reagent solution. By this
technique, the relationship between the area of spot and the logarithm of
the amount of insecticide present is linear for most of the insecticides
studied over the range 2.5 to 14 $\mu$g. per "spot".

In any discussion of possible "general" residue methods, reference must
also be made to the biological techniques which are popularly used "at least

to be able to show whether toxicologically significant residues are at all present": namely, cholinesterase techniques and bioassays.

Dimethoate inhibits cholinesterase about 100,000 times less than Phosdrin (DORMAL 1961). The inhibitor relationship between parathion and diazinon is 1 : 50, and after oxidation it is still 1 : 35 (DORMAL 1961). All possible intermediate stages may be observed for other insecticides. Such insignificance factors are undebatable from the aspect of analysis for enforcing legislation which is partly based on tolerance values of the order of tenths of a part per million. In view of facts in the pertinent literature and the techniques employed in laboratories, may I, however, be permitted a few words on the *cholinesterase technique*.

For carrying out extremely accurate measurements of cholinesterase activity and for studying its enzyme kinetics, the only method that can be considered is the manometric ("Warburg") technique. For measuring cholinesterase activity this technique is based on a procedure of AMMON (1934), which is still widely used today. The enzyme, buffer, and inhibitor are introduced in the side bulb of the vessel. After the temperature of the reaction mixture has reached that of the water bath and the taps have been closed, the side bulb is emptied into the substrate in the main compartment. After thirty minutes the degree of inhibition is calculated from the ratio of the carbon dioxide development of the inhibited mixtures to that of the control. This method and many of its modifications have a source of error which cannot be sufficiently emphasized, namely, the anticipated, uncontrollable inhibition that takes place before the mixture is added to the substrate. It is uncontrollable for the simple reason that it already takes place during the preparatory operations that precede the actual experiment. Fig. 4 demonstrates — for a constant inhibitor concentration — the influence of increasing incubation times of enzyme + inhibitor (para-oxon) on the carbon dioxide development at 37° C (FREHSE 1958, FREHSE and NIESSEN 1963).

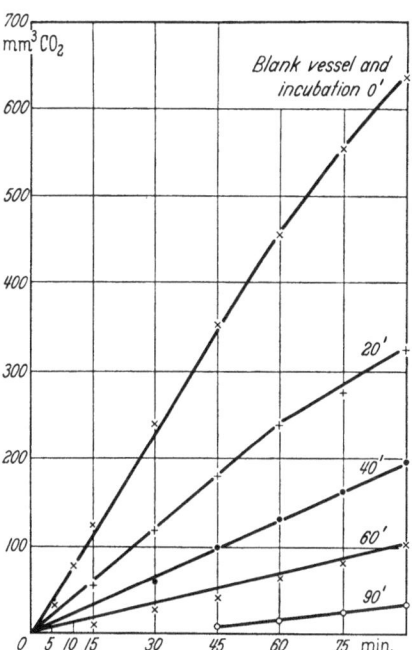

Fig. 4. Influence of increasing incubation times (0, 20, 40, 60, and 90 minutes) of enzyme + inhibitor on carbon dioxide development from bicarbonate buffer during measurement of cholinesterase inhibition. Substrate: acetylcholine chloride, $2.2 \times 10^{-1}$ molar solution; inhibitor: para-oxon, $3.3 \times 10^{-11}$ molar solution; enzyme: human blood plasma, diluted 1 : 15 (FREHSE 1958)

If controlled inhibiton conditions are not observed, differences amounting to several powers of ten may be simulated in the inhibitor concentration. We make allowance for these circumstances by pouring a mixture of inhibitor plus acetylcholine out of the side bulb into the enzyme at the beginning of the reaction, and by calculating the inhibiton from the initial velocities of the carbon dioxide evolution (Frehse 1958). Another possibility is to use vessels with two side bulbs. At the beginning of the experiment the inhibitor is added to the enzyme from one of the side bulbs, the mixture is incubated for a given time with the taps closed, and then the substrate is added (Frehse 1958). Afterwards the same procedure is followed as described in the method of Ammon (1933). Nesheim and Cook (1959) have concluded from similar studies that an incubation time of thirty minutes for enzyme plus inhibitor is sufficient for general analytical usage.

*Bioassay* with living organisms is also a non-specific method, and its possible uses for food control are regrettably often overrated. The absence of a mortality effect in a plant extract suspected of containing residues is rashly interpreted, by uncritical analysts, as proof of the absence of toxic residues in the bioassay, in fact even more so than the absence of an enzyme inhibition in cholinesterase assay. Such statements, however, are senseless from the analytical aspect. On the one hand, the susceptibility of the test organisms to the pesticidal active ingredients may vary by several powers of ten. Fig. 5 provides an example.

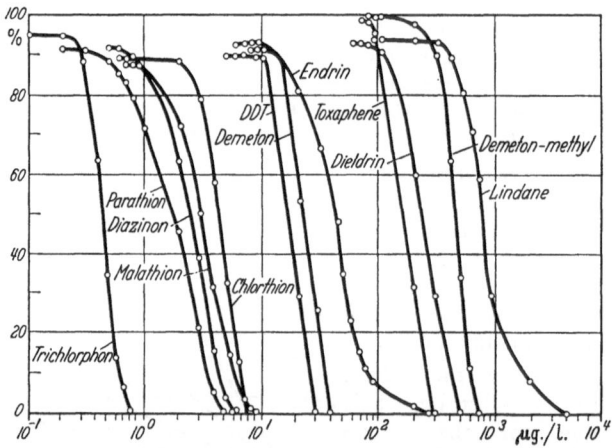

Fig. 5. Percentage of surviving *Daphnia* in relation to concentrations of several insecticides in water.
Redrawn from Bringmann and Kühn (1960)

The curves plot the percentages of test organisms *(Daphnia magna)* that survived the application against the concentration of active ingredient (Bringmann and Kühn 1960)[3]. Trichlorphon, the active ingredient of Dipterex and Dylox, is 1,800 times more toxic for *Daphnia* than is lindane. In other words, a mortality effect which indicates the presence of 2 p.p.m. of lindane could also be attained by 0.001 p.p.m. of trichlorphon (in Dipterex). Furthermore, such a test cannot be considered to represent a

---

[3] For similar results obtained with larvae of a pine cone moth see Merkel (1962).

measure of the general "toxicity" of a food. Table II illustrates that the toxicity of the active ingredients for *Daphnia* (listed as $LD_{50}$ in $\mu g./l.$) and warm-blooded animals (as $LD_{50}$ in mg./kg., rat) shows no parallels what-

Table II. *Toxicity of some insecticides against Daphnia magna and warm-blooded animals (albino rats)*

| Insecticide | Daphnia $LD_{50}$ in $\mu g./l.$ test solution[a] | Rats Oral $LD_{50}$ in mg./kg. body weight[b] |
|---|---|---|
| Dipterex | 0.4 | 625 |
| Parathion | 1.8 | 6.4 |
| Diazinon | 2.5 | 108 |
| Malathion | 3.0 | 1375 |
| Chlorthion | 4.2 | 625 |
| DDT | 15.0 | 250 |
| Demeton | 20 | 1.5—30 |
| Endrin | 39 | 17.8 |
| Toxaphene | 155 | 90 |
| Heptachlor | 200 | 100 |
| Thiodan | 220 | 45 |
| Dieldrin | 230 | 48 |
| Aldrin | 270 | 39 |
| Demeton-methyl | 440 | 120 |
| Lindane | 720 | 88 |

[a] Bringmann and Kühn (1960).
[b] For literature sources see Frehse and Niessen (1963).

soever. A falsification of results may also be incurred by choice of test organism. The active ingredient Bayer 29493, to which we again refer, is 200 times more toxic for mosquito larvae *(Aedes aegypti)* than its sulfoxide (Frehse and Niessen 1963), whereas its toxicity for the bean aphid *(Doralis fabae)* is about ten percent less than that of the sulfoxide (Schrader 1960).

There are two experiments which we consider deserve further consideration as methods of analysis for unknown residues. Areekul and Harwood (1962) use the following procedure. They determine the $LD_{50}$ for a certain test organism, e. g., *Artemia salina* (the brine shrimp), and apply this dosage to three other organisms which have no close biological relationship. In this way certain "fingerprints" are obtained for each insecticide by plotting the mortality range for these three organisms on a scale, as shown in Fig. 6 from an earlier study by these authors (Harwood and Areekul 1960). It ought to be possible to enlarge on this method so that it can be used for the identification of residues.

Another very interesting way of obtaining indications of the nature of an existing residue was recently pointed out by Phillips and co-workers (1962). Their method of *parallel screening* is a combination of bioassay, determination of organic chlorine, and measurement of cholinesterase inhibition. A lack of inhibition indicates that organophosphates are not present, and a lack of organically-bound chlorine naturally proves the absence of chlorinated insecticides. In addition to the data from each of these determinations, two additional information factors are obtained: the product of the $LD_{50}$ value and p.p.m. organic chlorine, and the ratio of the $LD_{50}$ to the micrograms required for 50 percent cholinesterase inhibition. The active ingredients that might be present can then be identified with a fair degree of accuracy from a table such as the one shown in Table III (Phillips *et al.* 1962).

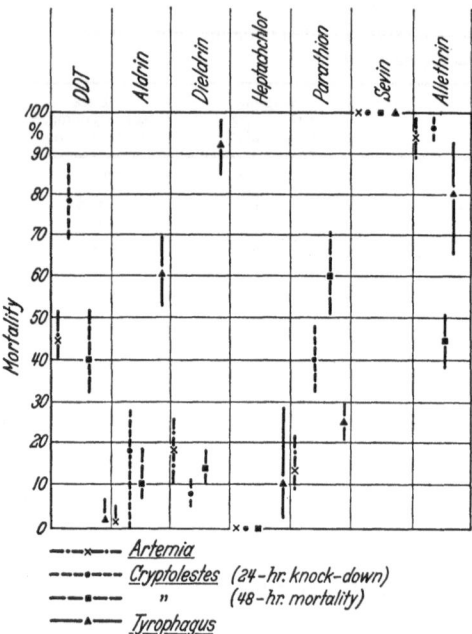

Fig. 6. Toxicity, for three organisms, of seven insecticides at the $LD_{50}$ level for drosophila. Redrawn from Harwood and Areekul (1960)

Comprehensive methods have occasionally been published, also, for the purpose of carrying out a pre-determination to establish whether any toxicologically significant residues are present. I should like to remind you of the study of Burchfield and Hartzell (1955). They investigated the inactivation of mosquito larvae by insecticides, as determined from the reaction of the larvae to a source of light. It was found that when the time required for inactivation of 50 percent of the larval population amounted to one hour and more, the absence of many insecticides in amounts of more than 0.1 p.p.m. was proved.

None of these tests alone represents a satisfactory solution of the problem. It ought to be possible, however, to arrive at a valid conclusion in many cases by means of suitable combinations of different methods.

I should like to close by discussing a problem of great importance for the whole field of residue analysis, including market controls. This is the problem of *accuracy in stating very small amounts of residue* in the range of the plant blank or control value. Several definitions have been proposed for uniform and general usage (Frehse and Niessen 1963, Frehse *et al.* 1962 a). A question inseparably linked with this problem is that of zero tolerance. At the meeting of the Entomological Society of America held on November 27, 1961 in Miami, Florida, Mr. W. V. McFarland of the U. S. Food and Drug Administration replied to the question "What do you mean by zero or 'no residue' and is there a difference?" as follows: "They are synonymous terms with respect to any specific pesticide. But as between individual pesticides zero or no-residue may mean something different. We recognize the impracticability of construing zero in the absolute sense. From a regulatory standpoint, a zero tolerance or no-residue clearance must be enforced just as any other tolerance by application of an analytical procedure which is satisfactorily sensitive in the light of what is known about the toxicity of the pesticide." SCHECHTER and WESTLAKE (1962) have outlined the problem in a manner which is somewhat more flattering for

Table III. *Toxicity and product values for organochlorine insecticides, and toxicity, $I_{50}$, and ratio $LD_{50}/I_{50}$ for organophosphate insecticides* (PHILLIPS et al. 1962)

| Organochlorine insecticides | | | | Organophosphate insecticides | | | |
|---|---|---|---|---|---|---|---|
| Compound | LD$_{50}$ ($\mu$g.) | Chlorine conversion factor | Product (LD$_{50}$, $\mu$g. Cl) | Compound | LD$_{50}$ ($\mu$g.) | I$_{50}$ ($\mu$g.) | Ratio |
| Ronnel . . | 0.02 | 0.333 | 0.0066 | Schradan . | 0.01 | 500 | 0.00002 |
| DDVP . . | 0.05 | 0.322 | 0.0166 | Phosdrin . | 0.01 | 4 | 0.0025 |
| Dieldrin . | 0.05 | 0.559 | 0.0279 | Methyl- | | | |
| Aldrin . . | 0.05 | 0.583 | 0.0293 | parathion | 0.05 | 10 | 0.005 |
| Dibrom . | 0.05 | 0.605 | 0.0305 | Dipterex . | 0.04 | 4 | 0.01 |
| Lindane . | 0.10 | 0.731 | 0.0731 | Malathion | 0.3 | 16 | 0.019 |
| Chlordane | 0.20 | 0.693 | 0.1386 | Demeton . | 0.75 | 30 | 0.025 |
| Thiodan . | 0.27 | 0.52 | 0.1404 | Ronnel . | 0.02 | 0.7 | 0.029 |
| Endrin . . | 0.30 | 0.559 | 0.1677 | Dibrom . | 0.05 | 1.0 | 0.05 |
| Heptachlor | 0.30 | 0.665 | 0.1995 | DDVP . . | 0.05 | 1.0 | 0.05 |
| BHC . . . | 1.0 | 0.731 | 0.731 | Guthion . | 1.0 | 14 | 0.071 |
| Toxaphene | 10 | 0.67 | 6.7 | Diazinon . | 0.18 | 0.4 | 0.45 |
| Chloro- | | | | Parathion. | 0.05 | 0.1 | 0.5 |
| benzilate | 32 | 0.218 | 6.72 | Trithion . | 0.8 | 1.0 | 0.8 |
| DDT. . . | 15 | 0.50 | 7.5 | Delnav. . | 10 | 4 | 2.5 |
| TDE . . . | 18 | 0.442 | 7.56 | Ethion . . | 5.6 | 0.5 | 11.2 |
| Methoxy- | | | | Co-Ral . . | 300 | 0.1 | 3000 |
| chlor . . | 500 | 0.308 | 154 | | | | |

we residue analysts: "The more capable the chemist, and the better his techniques, the harder it will be for him to prove that there is none of a compound $X$ in the sample analyzed."

Thus the following question arises: How sensitive is an analytical method when taking into consideration the usually unavoidable plant blank value? The sensitivity of a method is not an absolute measure because it can never be considered other than in relation to the amount and deviation of the plant blank value. We have therefore developed the following procedure for stating very small amounts of residue which approach the range of the blank value (FREHSE and NIESSEN 1963, FREHSE et al. 1962a).

In pre-tests, we determine the standard deviation of the blank ($\pm s$) for each analytical method and each plant material. In the course of time corresponding values also accumulate automatically in every residue laboratory. We refer to the value of $1.5\,s$ as "the lower limit of determination" on the grounds of the following deliberations.

If the analytical value $A$ and blank $B$ are each determined once in order to find a net value $N_{(1)}$, it follows from the Gaussian error-propagation law that the standard deviation $s_{N_{(1)}}$ of the net value is calculated to be

$$s_{N_{(1)}} = \pm \sqrt{s_A^2 + s^2}$$

where $s_A$ is the standard deviation of the analytical value and $s$ is the standard deviation of the blank. If the analytical value $A$ is of the same order as the blank $B$, $s_A$ can be approximately substituted by $s$. It follows that

$$s_{N_{(1)}} = \pm \sqrt{2\,s^2} = \pm 1.41\,s \approx 1.5\,s.$$

If the analytical value $A$ and blank $B$ are each determined twice in order to find a net value $N_{(2)}$, the standard deviation of this net value will be

$$s_{N_{(2)}} = \pm \sqrt{2\left(\frac{s}{\sqrt{2}}\right)^2} = \pm s.$$

As the distribution of error for blanks follows the normal Gaussian distribution, a statistical significance $P$ of 87 percent is obtained for a range of variation $\pm \lambda = \pm 1.5\,s_{N_{(2)}} = \pm 1.5\,s$. That means that the probability of mistaking a net value $N_{(2)}$ of 1.5 s p.p.m. for a fortuitous deviation of the blank amounts to $\frac{100-87}{2} = 6.5$ percent. Residues are detectable with a certain measure of probability also in the range of less than 1.5 s p.p.m. For the "lower limit of detection", we have introduced that value at which the probability of mistaking a net value for a fortuitous deviation of the blank is 30 percent. This limit is given by 0.5 s. However, as the determination of residues in the range between 0.5 s and 1.5 s is very vague, we do not express the residue in numerical values in these cases but instead we use the symbol $< 1.5$ s p.p.m. If the net value $N$ ist less than 0.5 s p.p.m., we refer to the residue as being "not detectable", a term which we abbreviate n. n. ("nicht nachweisbar"). We therefore express the residue as follows:

As the net value $N$, when $N$ is greater than 1.5 s (I)
As $< 1.5$ s, when $N$ is between 0.5 s and 1.5 s (II)
As n.n., when $N$ is less than 0.5 s (III)

Table IV shows the statistical distribution of the analytical results (net values $N_{(2)}$) in these three ranges for the calculation of the net value from two parallel analytical values and two parallel blank values.

Table IV. *Percentage of net values determined in ranges I, II, and III in relation to the quantity of residue actually present. Calculations are for 1.5 s as the lower limit of determination and 0.5 s as the lower limit of detection* (Frehse et al. 1962 a)

| Amount of residue present | I $N_{(2)} \geq 1.5\ s$ | II $0.5\ s \leq N_{(2)} < 1.5\ s$ | III $N_{(2)} < 0.5\ s$ |
|---|---|---|---|
| 0 | 6.5 | 24.3 | 69.2 |
| 0.5 s | 15.9 | 34.1 | 50.0 |
| 1.0 s | 30.9 | 38.3 | 30.9 |
| 1.5 s | 50.0 | 34.1 | 15.9 |
| 3.0 s | 93.5 | 5.9 | 0.6 |
| 4.5 s | 99.865 | 0.123 | 0.003 |

The time available does not permit me to enter into details of the mathematical argumentation which we have published in another paper (Frehse et al. 1962 a).

These proposed limits of determination and detection will fully satisfy practical requirements, for in view of the objectives that the residue analysis of pesticides is required to fulfill, the lower limit of determination should not be set too high. It is better to quote a residue value even though no more residue is actually present, rather than — when a high limit of determination is chosen — perhaps not to give a value which, from the toxicological aspect, would still be significant. On the other hand, we no longer express analytical results in terms of numerical values when it is not possible to give adequate statistical significance to these results. In addition, we avoid using the phrase "zero p.p.m.". Our definitions will, of course, only be expedient if the sensitivity of the applied analytical method is not restricted by other factors. It must be especially ensured that the analytical reaction is sufficiently sensitive in the range of $s$, and that the recoveries are large enough.

The residue analysts of all countries undoubtedly have hard times ahead of them. But they should not make their lives unnecessarily difficult by attempting to be more royalist than king. We have proved by carrying out three different methods of extraction on chopped, thoroughly mixed leaf material which contained radioactive residues, that parallel extractions may result in differences of ± 15 percent of radioactivity in the raw chloroform extracts (FREHSE and NIESSEN 1963). Therefore, why give residue values to two places after the decimal point? Of course, the legislators should also stop issuing tolerances such as 0.25 or 0.75 p.p.m.! Everyone knows that recoveries of 75 percent and more seem adequate for residue analyses (SCHECHTER and WESTLAKE 1962). Therefore, why should we not be able to use simplified methods for control purposes which neglect such metabolites that occur only to 10 to 20 percent, provided the other residues are adequately determined? Perhaps we should also endeavor to publish the results of our residue analyses in a form which provides the greatest possible abundance of information in the smallest possible space. We could make the fruits of our work more easily accessible to all, and also arrange for a better exchange of results. HOSKINS (1961) has found an interesting way of achieving this. He uses the quantities $-k_1$ and $k_2$ which are given by the equation for the break-down of residues:

$$\log r = -k_1 t + \log k_2 .$$

Table V gives an example from his comprehensive tables. These tables provide all information that is of importance for control organizations. The half-life ($t_{1/2}$) or residue-life 50 percent ($RL_{50}$) amounts, for example, to

$$RL_{50} = \frac{0.3}{k_1}$$

and the time required to reach the tolerance value amounts to

$$t_{tol.} = \frac{\log k_2 - \log tol.}{k_1} .$$

In closing, may I express the hope that experiences and modern laboratory techniques will, in the future, provide us with greater possibilities for carrying out analyses in the sub-microgram range. If we can dilute our

Table V. *Example for collection of data on amounts and persistence of insecticidal residues on plants*[a]

| Chemical | Formulation | Dosage/acre | Crop | Stage | I.D.[b] | $k_2$[c] | $k_1$[d] | $t_{1/2}$[e] | $t_{tol}$[f] | Reference and location |
|---|---|---|---|---|---|---|---|---|---|---|
| Parathion | 10% aerosol | 1 g./1,000 ft.³ | Lettuce | Mature | 3.8 | 15 | −0.063 | 4.8 | 18 | SMITH et al. 1954 (greenhouse plants) |
| | 15% WP | 1.5 lb./100gal. (1½gal./tree) | Peach Bark | 7 years old (early fruit) | 290 | 225 µg./cm.² | −0.0187 | 16 | | BOBB 1954 (Virginia) |
| | | 4.5 lb./100gal. (1½gal./tree) | | | 265 | 240 µg./cm.² | −0.0244 | 12 | | |
| | | 0.83 | Peaches | Immature | 0.5 (1 day) | 0.57 | −0.06 | 5 | | BRAID and DUSTAN 1955 (Ontario) |
| | | 1.66 | | mature | 1.17 (2 days) | 1.6 | | | | |
| | | 2 lb./100 gal. | | Green | 147 | 147 µg./fruit | −0.06 | 5 | | BRUNSON and KOBLITSKY 1952 (New Jersey) |
| | | | | Ripe | 0.6 (7 days) | 23 | −0.05 | 5 | 6 | |
| | 46.12% EC | 7 gal. of 0.03% spray per tree | | 10 days | 1.85 | 2.05 | −0.08 | 4 | 4 | DORMAL 1958 (Belgium) |
| | 46.7% EC | | | | 1.89 | 2.0 | −0.08 | 4 | 3 | DORMAL and MARTENS 1959 (Belgium) |
| | 15% WP | 1.5lb./100 gal. | | Ripe | 3.0 | 3.0 | −0.20 (0–3 days) | 1.5 | 2+ | FAHEY et al. 1952 (S. Indiana) |
| | | | | | | 0.9 | −0.044 (3–36 days) | 6.8 | | |
| | | 2 lb./100 gal. | | | 4.3 | 3.8 | −0.044 | | 13 | (N. Ohio) |
| | 25% WP | 0.2lb./100gal. | | 21 days | 0.7 | 0.75 | −0.08 | 4 | | USDA, July 9, 1959 (New Jersey) |
| | | 0.4lb./100gal. | | | 1.9 | 1.4 | −0.07 | 4 | 2 | |

[a] From HOSKINS (1961 a); compare with FREHSE and NIESSEN (1963).
[b] I. D. = initial deposit, p.p.m.
[c] Extrapolated deposit for zero time, p.p.m.
[d] Slope of time — log residue line.
[e] Half-life, days.
[f] Days to achieve tolerance value.

plant extracts instead of having to concentrate them, this will automatically lessen the problem of cleanup which today imposes limitations on most possibilities for developing general methods, and unfortunately still takes up the greater part of a residue analyst's valuable time.

## Summary

The residue chemist in the U.S.A. is today faced with the problem of having to consider 2,000 different tolerances for about 120 different active ingredients of pesticides when carrying out analytical work. A similar trend is also being encountered in European countries. Modern instrumentation has already opened up many new possibilities, and the residue methods have become more specific and more sensitive. Nevertheless, the task of the residue analyst is still divided: the "true" residue chemist who knows the history of his crops in full detail and applies specific methods must also be in a position to provide the "public chemist" with a recipe for an effective market control of crops with unknown history. This problem is especially acute in Europe where food legislation, with respect to pesticide residues, unfortunately is little uniform yet.

The possibilities and limitations of "general methods" for organophosphate, organochlorine, and carbamate pesticide residues are discussed. In addition, selected techniques are briefly presented: on the problem of tracing metabolites, and a critical study on the manometric measurement of cholinesterase inhibition. Further, attention is drawn to the "zero-tolerance" problem in connection with the difficulty of giving exact values for minute quantities of residues in the light of analysis sensitivity and standard deviations of plant blanks. As a solution to this problem appropriate definitions were worked out and suggested for general introduction. The paper closes with a few remarks on documentation of residue data and some future aspects of residue analysis.

## Résumé *

Aux Etats-Unis, le chimiste des résidus doit, à présent, faire face au problème soulevé par 2000 tolérances différentes, se rapportant à environ 120 matières actives de pesticides, lorsqu'il effectue son travail analytique. Une tendance similaire se développe aussi dans les pays européens. L'instrumentation moderne a déjà ouvert beaucoup de possibilités et les méthodes d'analyse des résidus sont devenues plus spécifiques et plus sensibles. Néanmoins, la tâche de l'analyste est encore partagée: le «véritable» chimiste des résidus qui connait, dans les moindres détails, l'histoire de ses récoltes et applique les méthodes spécifiques doit aussi être à même de fournir au «chimiste des services publics» une recette pour le contrôle effectif des marchés des récoltes dont les antécédents sont inconnus. Ce problème est particulièrement aigu en Europe où les réglementations régissant les denrées alimentaires présentent malheureusement peu d'uniformité quant aux résidus de pesticides.

---

\* Traduit par S. Dormal van den Bruel.

Les possibilités et les limitations des «méthodes générales» pour l'analyse des résidus de pesticides organo-phosphorés, organo-chlorés et des carbamates sont discutées. Des techniques choisies sont, en outre, présentées brièvement, notamment, sur le problème des métabolites traçant, et une étude critique sur la mesure manométrique de l'inhibition des cholinestérases. L'attention est aussi attirée sur le problème de la «tolérance zéro» en relation avec les difficultés d'évaluer avec précision les quantités infimes de résidus à la lumière de la sensibilité analytique et des déviations standard des témoins végétaux. Comme solution à ce problème, on a élaboré des définitions appropriées que l'on suggère comme introduction générale. L'étude se termine par quelques remarques sur la documentation relative aux données sur les résidus et la présentation de quelques aspects futurs de leur analyse.

## Zusammenfassung *

Der Rückstandsanalytiker steht in den USA heute vor dem Problem, 2000 verschiedene Toleranzen für etwa 120 verschiedene Wirkstoffe analytisch beherrschen zu müssen. Eine ähnliche Entwicklung bahnt sich auch in europäischen Ländern an. Moderne Instrumentation hat bereits viele neue Möglichkeiten eröffnet; die Rückstandsmethoden sind spezifischer und empfindlicher geworden. Trotzdem ist die Aufgabe des Rückstandsanalytikers noch geteilt: der „reine" Rückstandschemiker, der die Vorgeschichte seiner Proben im Detail kennt und spezifische Methoden anwendet, muß auch dem „öffentlichen Chemiker" ein Rezept für eine wirksame Marktkontrolle für Proben mit unbekannter Vorgeschichte an die Hand geben können. Dieses Problem ist in Europa besonders akut, wo die Lebensmittelgesetzgebung hinsichtlich der Pesticid-Rückstände leider noch wenig einheitlich ist.

Die Möglichkeiten und Grenzen von „Generalmethoden" für Organophosphor-, Organochlor- und Carbamat-Pesticid-Rückstände werden diskutiert. Darüber hinaus werden ausgewählte Arbeitstechniken kurz dargestellt: das Problem der Erfassung von Metaboliten und eine kritische Betrachtung der manometrischen Messung der Cholinesterase-Hemmung. Weiter wird das Problem der „Null-Toleranz" umrissen im Zusammenhang mit der Angabegenauigkeit kleinster Rückstandsmengen im Lichte der Analysenempfindlichkeit und der Standardabweichungen der Pflanzenblindwerte. Zur Lösung dieses Problems wurden geeignete Definitionen ausgearbeitet und zur allgemeinen Einführung vorgeschlagen. Die Arbeit schließt mit einer Betrachtung zur Dokumentation von Rückstandsdaten und einiger künftiger Aspekte der Rückstandsanalyse.

## References

AMMON, R.: Die fermentative Spaltung des Acetylcholins. Pflügers Arch. ges. Physiol. Menschen Tiere 233, 486 (1934).
ANONYMOUS: Wanted—general analytical procedure for detecting traces of several pesticides. Agr. Chemicals, Oct., 23 (1961).
AREEKUL, S., and R. F. HARWOOD: Experimental basis for estimating insecticides and acaricides by comparative bioassay. J. Econ. Entomol. 55, 894 (1962).

* Übersetzt vom Autor.

BERAN, F.: Das Problem der Pflanzenschutzmittelrückstände in europäischer Sicht. Pflanzenschutzberichte (Wien) 27, 11 (1961).

BRINGMANN, G., and R. KÜHN: Zum wasser-toxikologischen Nachweis von Insektiziden. Gesundheits-Ingenieur 81, 243 (1960).

BURCHFIELD, H. P., and A. HARTZELL: A new bioassay method for evaluation of insecticide residues. J. Econ. Entomol. 48, 210 (1955).

DORMAL, S.: Application de la mesure de l'inhibition des cholinesterases au dosage des residues de pesticides. Mededelingen Landbouwhogeschool Gent 26, 1508 (1961).

DURRENMATT, K.: Food regulations and international trade. Food Technol. 16 (Oct.), 91 (1962).

EICHENBERGER, J., and L. GAY: Zur semiquantitativen Bestimmung von Rückständen systemischer Insektizide in Pflanzenmaterial mit Hilfe der Papierchromatographie. Mitt. Gebiete Lebensmitteluntersuchung Hyg. (Bern) 51, 423 (1960).

ERWIN, E. R., D. SCHILLER, and W. M. HOSKINS: Preassay purification of tissue extracts by wax columns. J. Agr. Food Chem. 3, 676 (1955).

EVANS, W. H.: The paper-chromatographic separation and determination of chlorinated insecticide residues. Analyst 87, 569 (1962).

FADERL, N.: Methode zur Bestimmung von Mikromengen organischer Phosphorinsektizide. Mitt. Gebiete Lebensmitteluntersuchung Hyg. (Bern) 53, 154 (1962).

FREHSE, H.: Unpublished data 1958.

— Unpublished data 1961.

— Unpublished data 1963.

—, and H. NIESSEN: Die Analyse von Pflanzenschutzmittelrückständen. Z. Analyt. Chemie 192, 94 (1963).

— —, and H. TIETZ: Zur Anwendung des p.p.m.-Begriffs im Bereich sehr kleiner Pflanzenschutzmittel-Rückstände. Pflanzenschutz-Nachrichten „Bayer" 15, 113 (1962 a).

— — — Methode zur Bestimmung von Rückständen des Insektizids Lebaycid in pflanzlichem Material. Pflanzenschutz-Nachrichten „Bayer" 15, 152 (1962 b).

GETZ, M. E.: Six phosphate pesticide residues in, green leafy vegetables. Cleanup method and paper chromatographic identification. J. Assoc. Official Agr. Chemists 45, 393 (1962).

— The determination of organophosphate pesticides and their residues by paper chromatography. Residue Reviews 2, 9 (1963).

GUNTHER, F. A.: Instrumentation in pesticide residue determinations. Adv. Pest Control Research 5, 191 (1962).

HANCOCK, W., and E. Q. LAWS: The determination of traces of benzene hexachloride in water and sewage effluents. Analyst 80, 665 (1955).

HARWOOD, R. F., and S. Areekul: Identification of insecticides and acaricides by comparative bioassay. Science 131, 1369 (1960).

HOSKINS, W. M.: Final report on California's contributing project to Regional Project W-45. University of California, Berkeley 1961 a; FAO Plant Protection Bull. 9, 163 (1961 b).

LAWS, E. Q., and D. J. WEBLEY: The determination of organo-phosphorus insecticides in vegetables. Analyst 86, 249 (1961).

LEIB, E.: Pflanzenschutzmittel und Gesundheitsschutz. Nationale und Internationale Regelungen und Bestrebungen. Z. Pflanzenkrankh. (Pflanzenpath.) Pflanzenschutz 69, 641 (1962); Pflanzenschutzprobleme im Rahmen des Gesundheitsschutzes. Gesunde Pflanzen 15, 23 (1963).

LE MOAN, G.: Problemes analytiques poses par la detection et le dosage des residues de pesticides. Ann. Fals. Expt. Chim. 55, 63 (1962).

MAIER-BODE, H.: Die Insektizid-Rückstände bei der Kirschfruchtfliegenbekämpfung mit dem organischen Phosphorinsektizid Lebaycid. Anz. Schädlingskunde 35, 49 (1962).

MAJOR, A. JR., and H. C. BARRY: Screening method for pesticide residues on fruits and vegetables. J. Assoc. Official Agr. Chemists 44, 202 (1961).

McKINLEY, W. P., and J. H. MAHON: Identification of pesticide residues in extracts of fruits, vegetables, and animal fats. J. Assoc. Official Agr. Chemists 42, 725 (1959).

Merkel, E. P.: The toxicity of insecticides to larvae of *Dioryctria abietella* in laboratory screening tests. J. Econ. Entomol. 55, 682 (1962).

Mills, P. A.: Detection and semiquantitative estimation of chlorinated organic pesticide residues in foods by paper chromatography. J. Assoc. Official Agr. Chemists 42, 734 (1959).

Mitchell, L. C.: Separation and identification of chlorinated organic pesticides by paper chromatography. XI. A study of 114 pesticide chemicals: technical grades produced in 1957 and reference standards. J. Assoc. Official Agr. Chemists 41, 781 (1958).

Nesheim, E. D., and J. W. Cook: Cholinesterase inhibition method of analysis for organic phosphate pesticides: Effect of enzyme-inhibitor reaction time upon inhibition. J. Assoc. Official Agr. Chemists 42, 187 (1959).

Niessen, H., and H. Frehse: Eine infrarotspektroskopische Methode zur Bestimmung von N-Methylcarbamat-Rückständen in Pflanzen. Pflanzenschutz-Nachrichten „Bayer" 1963/4.

—, H. Tietz, and H. Frehse: Über das Vorkommen biologisch aktiver Umwandlungsprodukte des Wirkstoffs S 1752 bei der Anwendung von Lebaycid. Pflanzenschutz-Nachrichten „Bayer" 15, 129 (1962 a).

— — Papierchromatographische Trennung aromatischer Phosphorsäureester-Insektizide. J. Chromatog. 9, 111 (1962 b).

Phillips, W. F., M. C. Bowman, and R. J. Schultheiss: Estimation of insecticide residues in foods through parallel screening methods. J. Agr. Food Chem. 10, 486 (1962).

Schechter, M. S., and W. E. Westlake: Chemical residues and the analytical chemist. Anal. Chem. 34, 25 A (1962).

Schrader, G.: Die Entwicklung des Bayer-Präparates „S 1752". Höfchen-Briefe 13, 1 (1960).

Suter, R., R. Delley, and R. Meyer: Analysenmethoden einiger neuer Schädlingsbekämpfungsmittel. Z. analyt. Chemie 147, 173 (1955).

*Tagungsberichte:* 1. Internationale Arbeitstagung der Arbeitsgemeinschaft „Toxikologie von Pflanzenschutzmitteln". Nr. 42. Berlin: Deutsche Akademie der Landwirtschaftswissenschaften, 1962.

Tietz, H., and H. Frehse: Methode zur Bestimmung von Rückständen der systemischen Insektizide der Metasystox-Gruppe in pflanzlichem Material. Höfchen-Briefe 13, 212 (1960).

Unterstenhöfer, G., and H. Frehse: Wesen und Bedeutung der systemischen Wirkung von Insektiziden. Pflanzenschutz-Nachrichten „Bayer" 1963/4.

# Gas chromatography using an electron absorption detector

By

Lyle K. Gaston *

With 7 figures

## Contents

## I. Introduction

In 1952 James and Martin revived Martin's earlier discovery of gas-liquid partition chromatography. In 1959 Coulson applied gas chromatography to the separation of pesticides. As instruments and theory became more sophisticated, detectors advanced from the relatively insensitive automatic buret and katharometer to the extremely sensitive ion detectors. The most promising ion detector for pesticides is the electron affinity cell which forms the subject of this review.

## II. Electron affinity detector

In 1960 Lovelock described a variation of his original argon diode ion detector. Later (Lovelock 1961) he modified the geometry of the cell to that of two plane parallel plates (Fig. 1). The effluent from a gas chromatography column enters through a brass anode $A$. Usually a 100-mesh screen is placed near the anode to smooth out the gas flow through the cell. Attached to the brass cathode $C$ is a foil $R$ which contains a radioactive source of $\beta$-particles. Suitable sources are tritium ($\sim$ 100 mc.), strontium 90 ($\sim$ 10 mc.), promethium 147, or nickel 63. When there is only a non-elec-

* University of California Citrus Research Center and Agricultural Experiment Station, Riverside, California.

tron absorbing gas (nitrogen, helium) in the cell, the high energy $\beta$-particles (18 kev from tritium) produce positive ions and about a ten-fold increase of low energy electrons. By applying a potential to the electrodes,

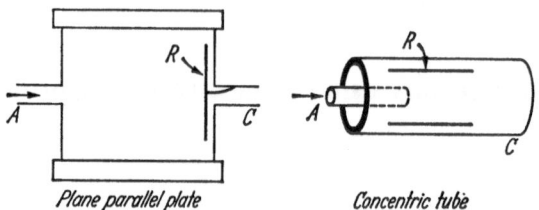

Plane parallel plate          Concentric tube

Fig. 1. Schematic drawing of two electron affinity cells. $A$ is the anode and the gas entrance, $C$ is the cathode, and $R$ is the radioactive beta-source

these electrons will migrate to the anode and thus establish a current. The applied potential determines the number of electrons collected. The voltage-current curve has the form shown later in Fig. 4. The part of the curve with zero slope (points 4 and 5 on the solid line) is called the plateau and represents the collection of all of the electrons in the cell. When a substance which can absorb these thermal electrons enters the cell, part of the electrons will be removed in the form of negative molecular ions. This decrease in the number of electrons causes a corresponding decrease in the current. The decrease in current is amplified and displayed on a strip chart recorder. For a more complete discussion of electron affinity detectors, LOVELOCK's original paper should be consulted (1961).

### III. Electron absorption by molecules

Substances can absorb thermal electrons by the two processes illustrated in Fig. 2. Process 1 represents the absorption of an electron by a substrate.

Absorption

$$AB + e^- \rightleftharpoons AB^- \pm \text{energy}$$

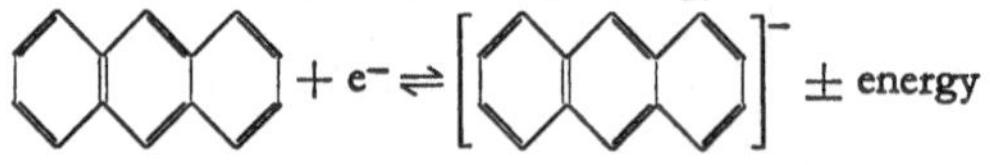

$$\text{(structure)} + e^- \rightleftharpoons \left[\text{(structure)}\right]^- \pm \text{energy} \tag{1}$$

Absorption and dissociation

$$AB + e^- \rightleftharpoons A \cdot + B^- \pm \text{energy}$$
$$CCl_4 + e^- \rightleftharpoons \dot{C}Cl_3 + Cl^- \pm \text{energy} \tag{2}$$

Fig. 2. The two processes by which compounds may absorb electrons

Process 1 occurs with compounds which have low-lying vacant atomic or molecular orbitals, either anti-bonding or non-bonding. The aromatic hydrocarbons form one class of compounds which have vacant low-lying antibonding orbitals.

Another class of compounds which can absorb thermal electrons via process 1 contains a highly dipolar functional group. This functional group may be an $\alpha$-dicarbonyl, an $\alpha$-dicyanide or their vinylogs and nitro groups. Other carbonyl groups such as ketones and esters have a very low affinity for electrons. In general, compounds which contain atoms with low lying vacant atomic (3*d* or 4*s*) or molecular orbitals can absorb electrons. Some examples of this are the phosphorus and sulfur containing compounds.

Process 2 represents the absorption of an electron followed by the dissociation of the substrate into a negative ion and a free radical. The efficiency of this process is a function of how easily a substrate molecule can absorb an electron and how easily it can dissociate. In general the anion formed is a halide ion. The absorption of an electron by an atom seems to be proportional to its polarizability which leads to the relative electron affinity series shown in Fig. 3 (LOVELOCK 1962). Another phenomenon is somewhat similar to a nuclear cross-section. The more electron absorbing atoms on the same central atom, the more efficient is the absorption process. The ease of dissociation depends upon the character of the radical; the more stable the radical is, the easier is the dissociation.

Most of the phosphorus insecticides fall into the first category whereas the halogenated ones fall into the second category.

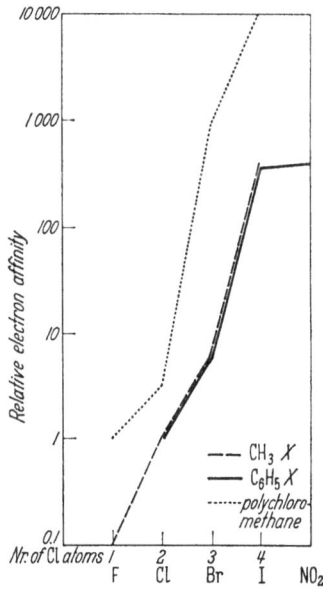

Fig. 3. Relative electron affinities of the polychloromethanes, methyl halides and halobenzenes

## IV. General comments on electron absorption detection in gas chromatography

The high sensitivity of the electron absorption detector requires that the operator exercise more care in its operation than is necessary for other less sensitive detectors.

The solvents used in the preparation of materials for electron absorption detection should be of high volatility and of the highest purity with respect to electron absorbing materials. The detection of highly electron absorbing substances at 0.1 p.p.m. in 10 $\mu$l. (microliters) is fairly easy. Fortunately the usual solvent impurities have electron absorption coefficients about $10^{-3}$ to $10^{-4}$ less than most of the chlorinated insecticides so that solvent purities of 99 percent to 99.9 percent are usually sufficient. Fractionally distilled hexane was found to be a satisfactory solvent. It has a low enough vapor pressure for standard solutions but adequate volatility so as not to interfer with early peaks in the gas chromatograph. Such solvents as c.p. benzene, spectro-grade isooctane and distilled xylene were found to interfer with the early eluding components. Chlorinated solvents, such as methylene chloride, are not useable as their high electron absorption coefficient gives rise to a very long solvent tail. Plastic ware, polyethylene, Tygon, etc., except

Teflon, should be avoided as they contain extractable materials which absorb electrons.

Cleanliness in all operations involving the compounds to be analyzed cannot be overemphasized. Syringes should be cleaned by drawing hexane through them with a vacuum. Special care should be exercised in cleaning the plunger and the outside of the syringe. We have found that syringes that were used in pesticide solutions of 50 $\mu$g./$\mu$l. and cleaned by the above procedure still gave the chromatogram of the original solution. These syringes were successfully cleaned in refluxing benzene. Needless to say, dilution pipets and volumetric flasks must also be cleaned thoroughly. Since in practice it is necessary to have on hand standard solutions whose concentrations will cover five powers of ten, the intercontamination of solutions is very easy.

With repeated use, various residues will collect on the foil on which the tritium is occluded. The rate at which the tritium foil becomes contaminated is a function of the amount and kind of material passed through the cell and the temperature of the cell. Contamination of a plane parallel-plate cell is characterized by a decrease in the standing current plateau and eventually by a noisy baseline. In the case of the concentric tube cell shown in Fig. 1, contamination of the foil is characterized by a decrease in the standing current plateau and eventually by a temporary increase in the total standing current following a peak. This corresponds to a deflection of the recorder below zero. The rate of contamination of the cell can be reduced by maintaining the cell at an elevated temperature. This minimizes the condensation of materials on the foil. The maximum temperature for a tritium source is 200° C. [1]. Even when the cell is kept at 200° C., the tritium source will become contaminated in time. Probably the liquid phase is responsible for some of the contamination as the cell becomes dirty faster at higher column temperatures. Oily samples seem to be the most efficient at contaminating the cell. Chlorophyll-containing materials are also very efficient.

The analysis of insecticides in the picogram (pg.; $10^{-12}$ g.) region demands that the solid support does not absorb or destroy any of the material. Our criterion for a good solid support is that support which when coated with a very thin film of solvent will give a reasonable response from 50 pg. of DDT with the very first injection. As yet we have not investigated all the supports available but Chromosorb-W [2] which was washed twice with hydrochloric acid for one day each, washed thoroughly with water and then silanized twice by BOHEMEN's (1961) procedure works very well. Columns made from this support showed no tailing at the two percent silicone oil loading and required no insecticide conditioning before use.

## V. Instrument parameters

A constant temperature column oven is necessary for the stability of an electron absorption detector used at maximum sensitivity. Constant temper-

---

[1] M. TAYLOR (1962) reported that a tritium source lost one percent of its activity per day at 250° C. and 0.1 percent of its activity per day at 200° C.

[2] Johns-Manville Co., 22 E 40th St., New York 16, New York.

ature means no on-off heater cycle as this causes a slight temperature fluctuation which in turn changes the bleed rate of the liquid phase. At maximum sensitivity this produces a sine wave for a base line. That the sine wave is actually due to a change in column bleed is indicated by a recent communication from the *Research Specialities Company* (1962). They report that the trimethyl silyl ether of steroidal alcohols can be detected by electron absorption. This indicates that the silicon atom with its low-lying $3d$ or $4s$ atomic orbitals (or possibly some low-lying molecular orbital between the silicon and oxygen atoms) can absorb a thermal election. Another feature of this bleeding pattern is that even though the column is purged at a much higher temperature, the liquid phase still continues to bleed at a reduced temperature.

In addition to a constant temperature, there should be no temperature gradients in the column oven. Temperature gradients decrease the column efficiency and therefore a longer analysis time is required for a given separation.

One of the most important operating parameters is the detector voltage. The detector voltage determines the value of the standing current. The amount of standing current determines the maximum amount of material per unit time which can pass through the detector.

For a plane parallel plate cell of the original LOVELOCK (1961) design a flow rate of 100 to 200 ml./min. is necessary to purge the cell of the eluded material and molecular ions. The family of curves in Fig. 4 represents the standing current as a function of the detector voltage for four different flow rates. It is noticed that the plateau, maximum standing current, is the same regardless of the flow rate; only the rate at which the plateau is reached is affected. A most interesting feature of this family of curves is the square on the 20 ml./min. flow rate curve. From the shape of this curve, the square corresponds roughly to circle 3 on the 100 ml./min.

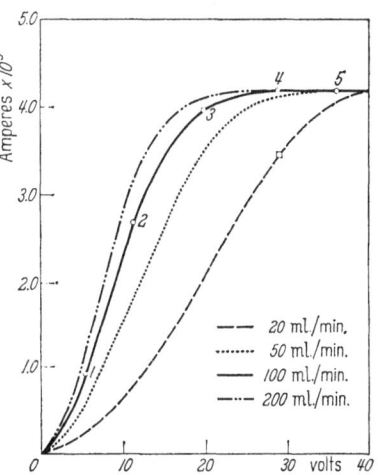

Fig. 4. The standing current as a function of the detector voltage at different flow rates for the plane parallel plate cell

flow-rate curve. In addition, this square can be made to correspond to circle 4 on the 100 ml./min. flow-rate curve. The significance of this point will become apparent after Fig. 5 is examined in detail.

Fig. 5 is a plot of peak area versus weight of aldrin at different detector voltages. The straight line 2 corresponds to the voltage (11.2 volts or 64 percent of the total standing current) of point 2 in Fig. 4. For this particular detector, aldrin has its greatest sensitivity (peak area per unit weight) at this voltage. Line 3 in Fig. 5 corresponds to point 3 in Fig. 4 (19.8 volts or 95 percent of the total standing current). The response is linear but the sensitivity toward aldrin is about $^2/_3$ that at 11.2 volts. Line *4* in Fig. 5 corresponds to point *4* in Fig. 4. It is now seen that the sensitivity is greatly decreased by approximately $^1/_5$, and the line curves badly. Since

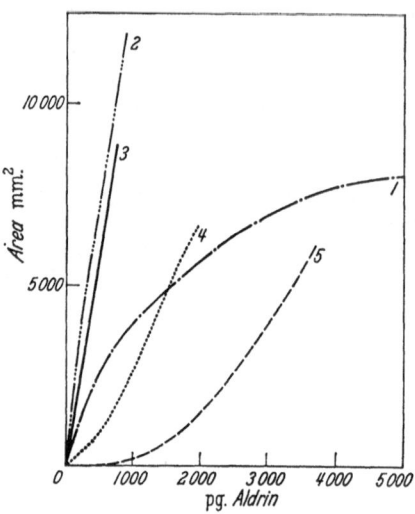

Fig. 5. The sensitivity of the plane parallel-plate cell at different voltages. The curve numbers refer to the numbered points in Fig. 4

this mode of operation is not very desirable, it is instructive to inquire into other ways of obtaining this condition. Clearly if one were to use a flow-rate of 20 ml./min. and a voltage corresponding to the square in Fig. 4 a sensitivity curve similar to line *3* in Fig. 5 would result. Now if the flow-rate is increased to 100 ml./min., point *4* of Fig. 4 will result and thus curve *4* in Fig. 5 will apply. Obviously a standard curve made at 29 volts and 20 ml./min. will be grossly in error when the flow-rate is increased to 100 ml./min. Curve *5* in Fig. 5 corresponds to point *5* of Fig. 4 (36 volts) and demonstrates the great loss in sensitivity when the detector is operated on its current-voltage plateau. The use of even higher detector voltages (60 volts) decreases the sensitivity even further, 30 ng. (nanograms, $10^{-9}$ g.) of aldrin being required for a reasonable size peak. Curve *1* in Fig. 5 corresponds to point *1* of Fig. 4. This voltage (5.5 volts) produces a sensitivity curve opposite of the two higher voltages. This is because the lower voltage establishes a smaller standing current and at the higher concentrations of substrate in the effluent gas stream almost all of this current is used up. As will become apparent later, the solution to this flow-rate problem is to add a purge gas to the cell at such a rate as to make the total flow through the cell equal to about 150 ($\sim$ 240 ml./min. at 200° C.) ml./min. at room temperature. For a constant small sample, 200 pg., the sensitivity as a function of the detector voltage goes through a maximum. The voltage corresponding to this maximum will be different for different compounds and different cells.

The current-voltage curve for the concentric tube detector cell shows only a slight flow dependence at low flow-rates (20 to 50 ml./min.). The sensitivity for a constant amount of substrate as a function of the detector voltage for two typical flow-rates is given in Fig. 6. Although a number of insecticides show a maximum in sensitivity at about 23 volts, in our hands this region has given a curved line for the weight-versus-area plot. In fact, at a flow-rate of 20 ml./min., all detector voltages gave curved lines for the area-weight plot. At a flow rate of 40 to 50 ml./min. and 90 volts, the area-versus-weight plot gave a fairly straight line. Other voltages at the higher flow rate gave curved lines for the sensitivity plot.

Previously for the plane parallel plate cell the effect of the flow-rate on the voltage-current curve was noted. The addition of a purge gas to the cell was suggested. A more quantitative method for examining the effect of

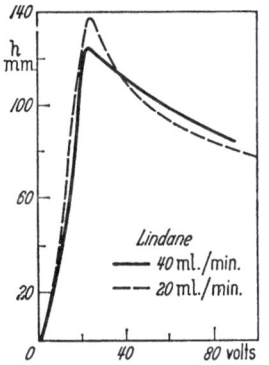

Fig. 6. The sensitivity (peak height) of the concentric tube cell as a function of voltage and flow rate

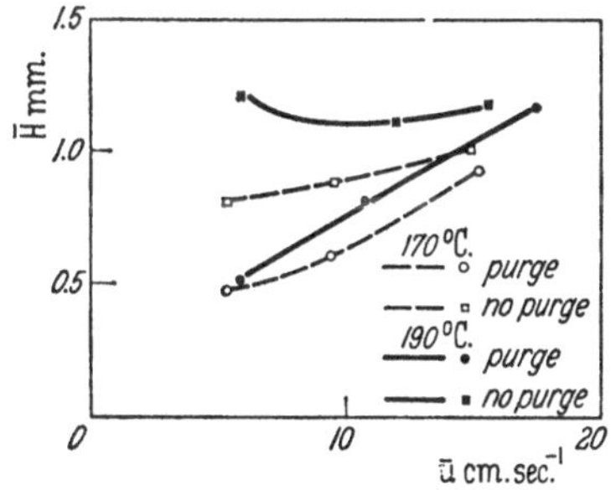

Fig. 7. A van Deemter plot using a plane parallel-plate cell showing the apparent increase in column efficiency with an added purge gas

a purge gas is to look at the average height equivalent of a theoretical plate, $\bar{H}$. In Fig. 7 is given a van Deemter plot, $\bar{H}$ vs the average velocity, $\bar{u}$, for one column at two temperatures and for only column flow and for an added purge gas. The number of theoretical plates was calculated from the following equation:

$$N = \frac{8\, t_R\, t_\beta}{\beta^2}$$

where $t_R$ is the time from injection to peak maximum, $\beta$ is the width of the peak at 0.368 of the peak height, and $t_\beta$ is the time from injection to the leading edge of the peak at 0.368 of the peak height (PURNELL 1962). At high linear velocities, the average HETP for only flow through the column approaches that for the addition of a purge gas. If a sufficiently high flow rate, 100 ml./min. minimum, can be obtained from the column itself, the addition of a purge gas will not reduce the time constant of the cell. At the low 10 to 50 ml./min. flow rates used with ⅛ inch columns, the addition of a purge gas to the extent that the total flow of gas through the cell is about 150 ml./min. can reduce the average HETP by as much as ½. Since the addition of a purge gas improves the column efficiency, the column itself must not be the limiting factor at low flow rates.

At the present time it is impossible to determine electron absorption coefficients for pesticides. One assumption is necessary in order to use WENTWORTH's (1962) method of determining electron absorption coefficients. That assumption is that all of the material injected into the chromatograph reaches the detector. Any losses of material in the injection block, column, column packing, or associated plumbing will decrease the electron absorption coefficient. A recent paper by BECKMAN (1963) indicates that these losses may be as much as 75 percent depending on the compound and the type of material it comes in contact. Relative electron absorption coefficients can be determined, but again these will reflect any losses in the gas chromatograph. Our own relative electron absorption coefficients are given in Table I. These were determined using a gas chromatograph equipped with a borosilicate injection block, borosilicate column, and three inches of stainless steel connecting tubing.

Table I. *Relative response of some insecticides to an electron affinity detector*

| Compound[a] | Relative peak area | Relative peak height |
|---|---|---|
| Aldrin . . . . . . . | 1.0 | 1.0 |
| Lindane . . . . . . . | 0.9 | 1.4 |
| Heptachlor . . . . . | 1.0 | 1.2 |
| Heptachlor epoxide . . | 0.8 | 0.7 |
| Dieldrin . . . . . . . | 1.0 | 0.6 |
| DDD (TDE) . . . . | 0.6 | 0.26 |
| DDT . . . . . . . . | 0.5 | 0.19 |
| Methoxychlor . . . . | 0.2 | 0.04 |

[a] Chemical names are given in Table II.

## VI. Literature review

GOODWIN et al. (1960 and 1961) reported a rapid screening test for pesticide residues in which they used an argon detector at a low voltage. By use of an acetone extraction and then a partitioning of the pesticide into hexane, they were able to analyze apples, broccoli, cabbage, carrots, grapes, lettuce, potatoes, swedes, tea, and tomatoes for the following insecticides with the lower limit of detection in p.p.m. (parts per million): lindane 0.1, heptachlor 0.1, aldrin 0.1, Telodrin 0.2, dieldrin 0.25, endrin 0.25, and DDT 1.0. Samples with residues of less than 0.1 p.p.m. were cleaned-up by column chromatography on alumina using hexane as the eluent. With this clean-up, lindane could be determined at the 30 p.p.b. (parts per billion) level. Insecticide recoveries were 80 to 90 percent at 0.5 to 20 p.p.m. Using a non-polar silicone column, they noted that grains contained a material which captured electrons and which had the same retention time as aldrin. By using a polar *ester* column, aldrin and the interfering materials were separated.

WATTS and KLEIN (1962) used a radium source in their electron affinity cell for nanogram quantities of chlorinated insecticides and a tritium source for picogram quantities. Collards and kale were analyzed for DDT (1 to 7 p.p.m.), aldrin (0.1 to 0.2 p.p.m.), and BHC (0.5 to 1.0 p.p.m.). They found that direct injection of uncleaned-up extracts gave poorly resolved peaks and led to a rapid contamination of the cell. Liquid-solid chromatography on Florisil eliminated most of the background interference. In addition, oily materials had to be partitioned between petroleum ether and acetonitrile before the Florisil chromatography.

MOORE (1962) used an argon detector with a radium source and either argon carrier gas and a cell potential of one to two volts or nitrogen carrier gas and a cell potential of nine volts and obtained a good response from Trithion, Methyl Trithion, phorate, malathion, DDVP, *p,p'*-DDT, *o,p'*-DDT, DDE, dieldrin, endrin, lindane, Bulan, and Prolan. Virginia pine foliage was analyzed for lindane and soil was analyzed for DDT. The lower limit of detection of lindane was found to be $10^{-12}$ g.

Table II. *Common or trademark and chemical names of pesticides mentioned in text*

| Common name | Chemical name |
|---|---|
| Aldrin. . . . . . | 1,2,3,4,10,10-hexachloro-1,4,4a,5,8,8a-hexahydro-1,4-*endo*, *exo*-5,8-dimethanonaphthalene |
| Bulan . . . . . . | 2-nitro-1,1-bis(*p*-chlorophenyl) butane [trademark name, Commercial Solvents Corp.] |
| BHC . . . . . . | mixture of stereoisomeric 1,2,3,4,5,6-hexachlorocyclohexanes |
| DDE . . . . . . | 1,1-dichloro-2,2-bis(*p*-chlorophenyl) ethylene |
| *o,p'*-DDT . . . . | 1,1,1-trichloro-2-*o*-chlorophenyl-2-*p*-chlorophenylethane |
| *p,p'*-DDT . . . . | 1,1,1-trichloro-2,2-bis(*p*-chlorophenyl) ethane |
| DDVP . . . . . | *O,O*-dimethyl-2,2-dichlorovinyl phosphate |
| Dieldrin . . . . . | 1,2,3,4,10,10-hexachloro-*exo*-6,7-epoxy-1,4,4a,5,6,7,8,8a-octahydro-1,4-*endo*, *exo*-5,8-dimethanonaphthalene |
| Endrin . . . . . | 1,2,3,4,10,10-hexachloro-*exo*-6,7-epoxy-1,4,4a,5,6,7,8,8a-octahydro-1,4-*endo*, *endo*-5,8-dimethanonaphthalene |
| Heptachlor. . . . | 1,4,5,6,7,8,8-heptachloro-3a-4,5,5a-tetrahydro-4,7-*endo*-methanoindene |
| Heptachlor epoxide | 1,4,5,6,7,8,8-heptachlor-2,3-epoxy-2,3,3a,7a-tetrahydro-4,7-methanoindene |
| Lindane . . . . . | gamma isomer of 1,2,3,4,5,6-hexachlorocyclohexane |
| Malathion . . . . | *O,O*-dimethyl *S*-(1,2-dicarbethoxyethy) phosphorodithioate |
| Methoxychlor . . | 1,1,1-trichloro-2,2-bis(*p*-methoxyphenyl) ethane |
| Methyl Trithion . | *O,O*-dimethyl *S*-[(*p*-chlorophenylthio) methyl] phosphorodithioate |
| Phorate . . . . . | *O,O*-diethyl-*S*-ethylthio methylphosphorodithioate |
| Prolan. . . . . . | 2-nitro-1,1-bis(*p*-chlorophenyl) propane [trademark name, Commercial Solvents Corp.] |
| TDE . . . . . . | 1,1-dichloro-2,2-bis(*p*-chlorophenyl) ethane |
| Trithion . . . . . | *S*-(*p*-chlorophenylthiomethyl) *O,O*-diethyl phosphorodithioate [trademark name, Stauffer Chemical Co.] |

A. TAYLOR (1962) used an electron affinity detector to analyze breast muscle and liver for dieldrin, heptachlor epoxide, and lindane. MATTICK (1963) used an electron affinity detector to determine endrin residues on cabbage at a minimum of 20 p.p.b. Fortified samples gave recoveries of $98 \pm 7$ percent. The standard curve was reported to be linear between zero and 100 ng. with a standard error of 0.7 ng.

## Summary

Gas Chromatography followed by electron absorption is a very sensitive tool for the detection of chlorinated (10—100 pg.) or phosphorothionate (0.1—10 ng.) insecticides. The response of the detector is dependent on the current flow through the cell and the maintainance of a constant and known current requires considerable operator care [3]. Changes in the current may be caused by a change in voltage, flow rate and/or materials

---

[3] A recent article by LOVELOCK, Anal. Chem. **35**, 474 (1963), should be read thoroughly and appreciated.

collecting on the radioactive source. The fluctuations due to the flow rate can be minimized by using a high flow (150 ml./min.) part of which may be a purge gas added directly to the cell. In general, uncleaned-up samples will deposit materials on the radioactive source which in time will necessitate cleaning the source. In some cases one injection of an extract of natural materials will destroy the source's efficiency. With simple clean-up procedures, most chlorinated insecticides can be conveniently determined down to 0.1 p.p.m.

### Résumé *

La chromatographie en phase gazeuse suivie par l'absorption d'électrons est un outil très sensible pour la détection des insecticides organo-chlorés (10—100 pg) ou organo-thiophosphorés (0.1—10 ng.). La réponse du détecteur dépend du débit du courant qui traverse la cellule et l'opérateur doit avoir grand soin de maintenir un courant constant de densité connue. (On lira attentivement un récent mémoire de Lovelock: Anal. Chem. 35, 474 [1963] à ce sujet.) Des variations de courant peuvent être provoquées par un changement de tension, de débit de gaz et aussi par les substances qui se rassemblent sur la source radioactive. Les fluctuations dûes au débit gazeux peuvent être réduites par l'utilisation d'un débit élevé (150 ml par mn) dont une partie peut être dûe à un gaz épuré ajouté directement à la cellule. En général, des échantillons non épurés déposent des matières sur la source radioactive et rendent à la longue son nettoyage nécessaire. Parfois une seule injection d'un extrait de substance naturelle annihile l'efficacité de la source. La plupart des insecticides organo-chlorés peuvent être commodément déterminés jusqu'à 0,1 ppm avec des procédés simples de purification.

### Zusammenfassung **

Gaschromatographie mit nachträglicher Elektronenabsorption ist ein äußerst sensitives Mittel zur Bestimmung von chlorierten (10—100 pg.) oder Thiophosphorsäureester-Insecticiden (0,1—10 ng.). Das Ansprechen des Detektors hängt vom Stromfluß durch die Zelle ab. Die Aufrechterhaltung eines konstanten und bekannten Stromes erfordert beträchtliche Sorgfalt vom Ausführenden [4]. Stromänderungen können durch Änderung der Spannung, der Durchflußmenge und/oder Material, das sich an der radioaktiven Quelle sammelt, verursacht werden. Die Schwankungen, die auf die Durchflußmenge zurückzuführen sind, können verringert werden durch Erhöhung der Durchflußmenge (150 ml/min.), die teilweise aus einem Spülgas bestehen kann, das der Zelle direkt zugefügt wird. Im allgemeinen scheiden nicht aufgearbeitete Muster Material an der radioaktiven Quelle aus, so daß diese von Zeit zu Zeit gereinigt werden muß. In gewissen Fällen kann die Wirksamkeit der Quelle durch eine Injektion eines Extrak-

---

* Traduit par R. Mestres.
** Übersetzt von H. Martin.
[4] Ein wertvoller, neuer Artikel von Lovelock, Anal. Chem. 35, 474 (1963) sollte eingehend gelesen werden.

tes aus natürlichem Material zerstört werden. Mit einfachen Aufarbeitungs-
methoden lassen sich die meisten chlorierten Insecticide bis zu 0,1 ppm
leicht bestimmen.

## References

BECKMAN, H., and A. BEVENUE: The effect of the column tubing composition on
the recovery of chlorinated hydrocarbons by gas chromatography. J.
Chromatogr. 10, 231 (1963).

BOHEMEN, J., S. H. LANGER, R. H. PERRETT, and J. H. PURNELL: A study of the
adsorptive properties of firebrick in its relation to its use as a solid support in
gas-liquid chromatography. J. Chem. Soc. 2444 (1960).

GOODWIN, E. S., R. GOULDEN, A. RICHARDSON, and J. G. REYNOLDS: The analysis
of crop extracts for traces of chlorinated pesticides by gas-liquid partition
chromatography. Chem. & Ind. 1220 (1960).

— —, and J. G. REYNOLDS: Rapid identification and determination of residues
of chlorinated pesticides in crops by gas-liquid chromatography. Analyst 86,
697 (1961).

JAMES, A. T., and A. J. P. MARTIN: Gas-liquid partition chromatography: the
separation and micro-estimation of volatile fatty acids from formic acid to
dodecanoic acid. Biochem. J. 50, 679 (1952).

LOVELOCK, J. E., and S. R. LIPSKY: Electron affinity spectroscopy — A new method
for the identification of functional groups in chemical compounds separated by
gas chromatography. J. Amer. Chem. Soc. 82, 431 (1960).

— — Ionization for the analysis of gases and vapors. Anal. Chem. 33, 162 (1961).

— — Analysis by gas phase thermal electron absorption. 1st Ann. Pacific Meeting
Appl. Spectroscopy and Anal. Chem., Pasadena, Calif., Oct. 18—19, 1962.

MATTICK, L. R., O. L. BARRY, F. M. ANTENUCCI, and A. W. AVENS: The
disappearance of endrin residues on cabbage. J. Agr. Food Chem. 11, 54 (1963).

MOORE, A. D.: Electron capture with an argon ionization detector in gas
chromatographic analysis of insecticides. J. Econ. Entomol. 55, 271 (1962).

PURNELL, H. J.: Gas Chromatography. New York and London: Wiley 1962.

Research Specialties Co.: Chromatofacts. Sept.-Oct. 1962.

TAYLOR, A.: A rapid determination of seed-dressing insecticide residues in animal
relicta. Analyst 87, 824 (1962).

TAYLOR, M. P.: Possible radiation hazards arising from the use of radioactive
detectors in gas chromatography. J. Chromatogr. 9, 28 (1962).

WATTS, J. O., and A. K. KLEIN: Determination of chlorinated pesticide residues
by electron capture gas chromatography. J. Assoc. Official Agr. Chemists 45,
102 (1962).

WENTWORTH, W. E., and R. S. BECKER: Potential method for the determination
of electron affinities of molecules: Application to some aromatic hydrocarbons.
J. Amer. Chem. Soc. 84, 4263 (1962).

# Quantitative determination of pesticide residues by electron absorption chromatography: Characteristics of the detector

By

S. J. CLARK *

With 6 figures

## Contents

## I. Introduction

Practical exploitation of electron absorption detection followed rapidly upon the original announcement of the technique (LOVELOCK and LIPSKY 1960) and it is now employed in the gas chromatographic determination of pesticide residues (GOODWIN, GOULDEN, and REYNOLDS 1960, CLARK 1960, WATTS and KLEIN 1961), lead alkyls in gasoline (LOVELOCK and ZLATKIS 1961, DAWSON 1963) and in air pollution studies (BELLAR and SIGSBY 1963, DARLEY, KETTNER, and STEPHENS 1963). In addition, LOVELOCK (1961 a, 1962, 1963) has considered the relationship between electron absorption and biological activity, and WENTWORTH and BECKER (1962) have described the

---

\* Jarrell-Ash Company, Newtonville, Massachusetts.

use of an electron absorption detector for the measurement of electron affinities.

As with all novel approaches to analytical determinations, some problems have arisen in the application of the technique to pesticide residue analysis: The selectivity of the method has necessitated revision of clean-up techniques; for some matrices, considerably less clean-up is required; for others, older procedures must be modified for best results. The extremely high sensitivity of the electron absorption detector has exposed inadequacies in columns and injection systems in terms of loss or destruction of sample and has increased the problem of preparing stable high temperature columns. Finally, it is necessary to know the effect of experimental variables upon detector performance in order that the device may be operated under optimum conditions.

Little has yet been published on detector characteristics with respect to quantitative performance. LANDOWNE and LIPSKY (1962) reported on the response of the detector to alkyl halides, but failed to present a logical relationship between detector response and experimental variables. A theoretically derived relationship between response and sample concentration has been proposed (COULSON 1962) but some of the assumptions upon which the derivation is based are questionable.

The work reported in this paper forms part of a continuing program of study of electron absorption techniques and is concerned with detector performance. The investigation, for the most part, has been directed to the elucidation of the behaviour of the detector in the determination of chlorinated pesticides, but the findings are generally applicable to other types of strongly absorbing molecules.

## II. Experimental conditions

### a) Apparatus

A Jarrell-Ash Model 700 Universal Chromatograph was used throughout the investigation. The detector (Cat. No. 26-755) followed the design of LOVELOCK (1961 b) having plane-parallel geometry and an electrode spacing of 1.5 cm. The tritium source was of such activity as to furnish a saturation current in the range 5.0 to $9.0 \times 10^{-9}$ ampere.

For pulsed operation, a Simpson Electric Company Model 2620 Pulse Generator was employed. The generator was modified to include a d. c. restorer circuit in order to maintain a stable recorder base-line. The circuit consisted of a silicon diode connected in parallel with an 18 kilohm resistor between the output terminal of the generator and ground.

An auxiliary supply of carrier gas was connected between column and detector by means of a Swagelok "Tee" fitting.

Clean, dry nitrogen (Linde liquid nitrogen, or Air Reduction Company prepurified grade) or a mixture of 90 percent argon and 10 percent methane (Matheson Company) were used as carrier gases.

Columns were constructed from stainless steel, glass or Teflon and were packed with 80—90 mesh Anakrom AS (Analytical Engineering Laboratories) coated with 3 to 10 percent of the appropriate stationary phase.

## b) Reagents

For detector studies, purified materials were used; gas chromatographic analysis showed purity to be 99.5 percent or better for all compounds employed. Sensitivities for the listed pesticides were determined using commercial samples in order to obtain values representative of those likely to be found in practice. Purities of these materials ranged from 85 to 95 percent.

## c) Precision of measurements

Statistical analysis of all results used in this investigation showed precision to be ±5.6 percent at the 95 percent confidence level. The major contributions to lack of precision, so far observed, arise from the following sources:

*1. Injection of sample.* Precision of injection of from one to six microliters of solution, using a Hamilton microliter syringe, was found to be ±2 to 3 percent.

*2. Column stability.* Changes in column bleed rate accounted for most of the remainder of the error. It was found, in general, that as much as seven days conditioning was required before many columns were sufficiently stable for use. In general, the less polar phases, e. g. silicone grease, SE-30, SE-52, provided the most stable columns. Material such as XE-60 were intermediate in stability and the more polar phases, e. g. QF-1, the polyesters, and the polyalkalene oxides, were rather unstable.

Even under best conditions, some degradation of chlorinated materials occurred. This resulted in a slow build-up of degradation products on the column which eventually broke through, causing a decrease in standing current in the detector. In order to minimize trouble from this source, the following routine was established: At the end of the day, the column was disconnected from the detector and a Swagelok plug attached to the fitting remaining connected to the detector. Carrier gas from the auxiliary line was passed through the detector overnight in order to maintain it in operating condition. The column was raised to its maximum operating temperature and was purged overnight with carrier gas. Each morning, the column was reconnected to the detector and analytical conditions re-established. The same series of operations could be more conveniently carried out by using a suitable high temperature valve attached to the column outlet.

*3. Column temperature control.* Fluctuations in column temperature will obviously affect column bleed rate. Control of column temperature to ±0.1° C. was found adequate for most purposes.

*4. Flow rate.* At a flow-rate of 100 ml. per minute, variations of approximately ±2 percent can be tolerated. The Moore differential flow controller used in conjunction with a good needle valve is more than adequate to provide this precision.

*5. Detector temperature.* Detector temperature was maintained at the required value ±1° C. No changes in sensitivity could be detected for fluctuations of this magnitude.

## III. Results and discussion

The detector depends for its operation upon the reaction of sample molecules with electrons to form negative ions. The consequent reduction in negative ionic mobility is detected either directly, in the pulsed mode of operation, or through ion recombination, in the d. c. mode.

The main processes occurring in the detector with pure carrier gas flowing may be summarized as follows:

1. Ionization of carrier gas molecules to produce positive ion-electron pairs.
2. Separation of ion pairs.
3. Collection of charge carriers.

When sample gas enters the detector, further reactions occur:

4. Electron attachment by sample molecules to form negative ions.
5. Recombination of positive and negative molecular ions.

For a given compound, the response of the detector depends upon the extent to which negative ions are formed and lost; any change in conditions affecting these processes will lead to a change in sensitivity.

### a) Applied potential

Ionization of the carrier gas occurs randomly throughout the ionization chamber and the resulting ions have a random distribution of velocity and direction. Thus, in a d. c. polarized cell, the form of the current-voltage relationship should be gaussian but in practice the curve is distorted by space charge effects. A comparison of calculated and experimental standing current curves at a single flow-rate is shown in Fig. 1.

At very low applied potentials, separation of the positive ion-electron pair is less likely to occur and recombination can take place. Response to sample entering the detector is a function of free electron concentration in the ionisation chamber, hence sensitivity would be expected to increase in proportion to the standing current as potential is increased, reaching a maximum at saturation. The third curve in Fig. 1 illustrates the relation between voltage and response; it is seen that, in fact, response reaches a maximum at approximately that voltage theoretically required for saturation. The presence of sample molecules apparently reduces the effect of the

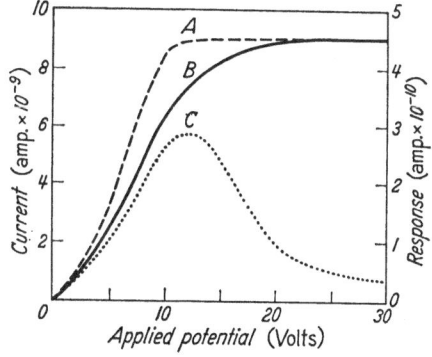

Fig. 1. Relationship between applied potential and response. Detector temperature: 220° C; carrier gas: nitrogen 40 ml./sin.; sample: $5 \times 10^{-11}$ g. lindane. *A*: theoretical current-voltage curve, *B*: experimental current-voltage curve, and *C*: response curve

positive ion space charge, through molecular ion recombination, allowing maximum response at a voltage lower than would have otherwise been expected.

At potentials greater than that required for maximum response, the charge carriers in the chamber are further accelerated, thus reducing the probability both of electron absorption and of ion recombination. In this region, sensitivity is approximately inversely proportional to the square of the applied potential.

### b) Pulsed operation

In this mode of operation, highest sensitivity is attained when the maximum time is provided for reaction to occur; that is, using a pulse of minimum width at a low repetition rate. At pulse widths and repetition rates below a certain value, electrons are lost through recombination, and at the walls of the ionization chamber. Thus, response will exhibit a maximum for certain experimental conditions. A pulse repetition rate of 10 kilocycles per second and a width of one microsecond were found to give maximum sensitivity. Response is decreased at higher or lower pulse repetition rates. At a repetition rate of 10 kilocycles per second, pulse width has no effect on sensitivity in the range 1 to 3 microseconds. The detector is insensitive to pulse amplitude in the range 35 to 90 volts; below 35 volts sensitivity is decreased. Pulse shape does not appear to be critical; however, a generator providing reasonably square pulses is recommended since the limits beyond which pulse shape affects response have not yet been determined.

### c) Carrier gas flow rate

Fig. 2 shows current-voltage relationships for a number of flow-rates. A change in gas flow-rate will affect the drift velocity of ions in the chamber and the geometry of the chamber is such that positive ions move in the

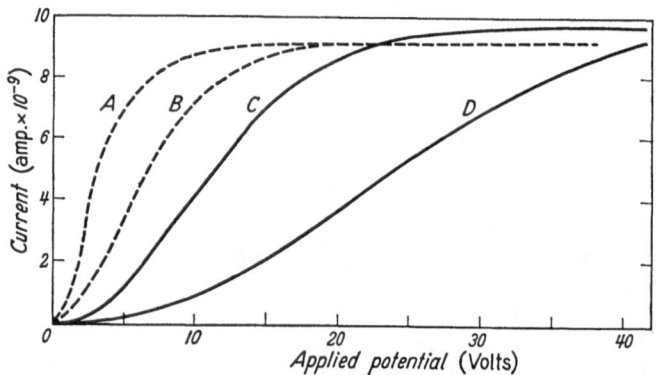

Fig. 2. Effect of temperature and flow rate upon standing current. Carrier gas: nitrogen

|   | Detector temperature, ° C. | Flow rate, ml./min. |
|---|---|---|
| A | 220 | 160 |
| B | 220 | 40 |
| C | 120 | 160 |
| D | 120 | 40 |

same direction as the carrier gas stream. Thus an increase in flow-rate will reduce the randomness of ionic velocity and saturation will occur at a lower applied potential. For each flow-rate, a response curve can be constructed exhibiting maximum sensitivity at approximately theoretical saturation potential for that flow-rate. The relationship between flow-rate and potential for maximum response is shown in Fig. 3. Sensitivity at maximum response is constant over the flow-rate range 25 to 200 ml. per minute and occurs always at the same value of standing current. Thus, provided that voltage is adjusted to compensate for changes in flow-rate, sensitivity remains constant. However, it is obviously essential for accurate work to employ precise flow control and if temperature programmed operation is contemplated, some form of differential flow regulation is mandatory. The deleterious effect of poor flow control can be alleviated to some extent by operating the detector at high flow-rates, where it is less sensitive to flow changes. If the column must be operated at low flow-rates, additional carrier gas may be introduced at a point between the column and the detector.

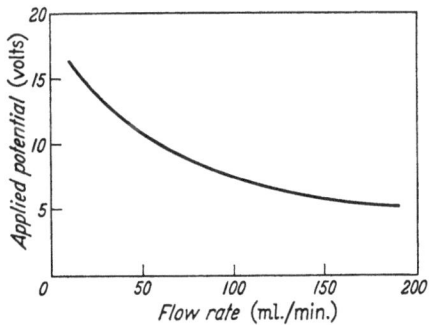

Fig. 3. Relationship between potential for maximum response and flow rate. Carrier gas: nitrogen

Operated in the pulsed mode, the detector is insensitive to flow-rate changes over the range 40 to 200 ml. per minute. Here, the pulse duration is too short to accelerate positive ions and only electrons are collected. Hence, changes in the motion of positive ions will not be observed.

### d) Detector temperature

Movement of the ions in the chamber is temperature dependent so that changes in detector temperature in a d.c. polarized system will affect potential for maximum response in much the same way as do changes in flow-rate. The effect upon standing current was illustrated in Fig. 2. For pulsed operation, little change in standing current was observed over the range 110 to 225° C. Presumably, electron mobility is so high that it is relatively unaffected by temperature changes.

All of the reactions occurring in the ionization chamber are temperature dependent so that a temperature effect would be expected in addition to that on ionic mobility. The series of reactions occurring in the detector is far too complex to allow, for the present, quantitative evaluation of the temperature dependence of each. It seems likely, however, that apart from the electron attachment reaction, temperature coefficients will be of approximately the same magnitude for all components.

Dependence of sensitivity upon temperature for two different types of strongly capturing compounds is illustrated in Fig. 4. It will be seen that change of response with temperature is strongly dependent not only upon compound type but also upon the mode of operation of the detector. Temperature response curves for the less strongly capturing compounds anthracene and 1,2,3,4-tetrachlorobenzene were also measured. The curves for anthracene have the same general form as those for diethylmaleate, whereas 1,2,3,4-tetrachlorobenzene exhibits yet a third type of behaviour. Here sensitivity decreases by approximately 20 percent over the range

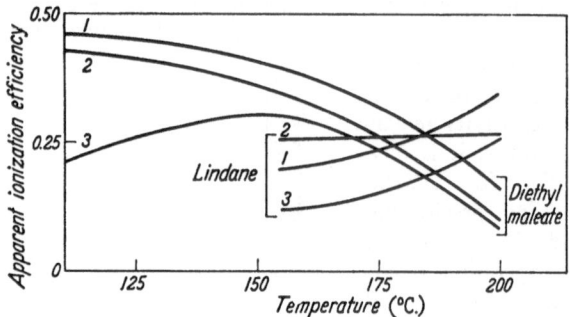

Fig. 4. Dependence of sensitivity upon temperature. Carrier gas flow rate 150 ml./min. *1*: D.C. argon-methane; *2*: Pulse, argon-methane; and *3*: D.C. nitrogen

110 to 160° C., remaining almost constant thereafter. For both compounds sensitivities decrease in the order d.c. — argon-methane; pulse; d.c. — nitrogen.

It is evident that the temperature dependence of the detector is extremely complex and much more experimental work would be required before any general statement could be made relating, for example, electron temperature to sensitivity. However, several important practical conse-quences emerge: Firstly, the detector must be maintained at a fixed temperature independent of the remainder of the instrument. Secondly, comparison of results from different sources will be fruitless unless the temperature dependence of the system is established, or unless experiments are universally conducted at a fixed temperature. For pesticide residue work it is suggested that the detector be maintained at 210° C.; this value is sufficiently high to minimize contamination from column bleed yet low enough so that loss of tritium from the radioactive source is negligible. Thirdly, the effect may be used to advantage either for qualitative work or to select conditions such that interference to the compound sought is minimized. Finally, caution must be exercised in comparing results obtained with d.c. operated detectors having differing geometries. Here, the distri-bution of electron energies is affected not only by temperature changes, but by the non-uniformity or otherwise of the potential gradient across the ionization chamber.

### e) Sensitivity of electron absorption measurement

The ultimate sensitivity of any ionization detector is a function of fluctuations in the detector standing current. For an electron absorption detector having a saturation current of $1 \times 10^{-8}$ ampere, the noise asso-ciated with the ionization process is approximately $5 \times 10^{-4}$ ampere and the corresponding limit of detection approximately $5 \times 10^{-19}$ mole per second. In practice, noise from other sources greatly exceeds that from the ionization process so that in a practical system, a noise level of the order of $1 \times 10^{-12}$ ampere may be expected. Thus a practical limit of detection is approximately $1 \times 10^{-17}$ mole per second and this will be further reduced since ionization of the sample is rarely 100 percent efficient. Table I lists

Table I. *Sensitivities for some common chlorinated pesticides* [a]

| Pesticide | Apparent ionization efficiency ($E$) | Limit of detection mole/sec. | Acute oral toxicity ($LD_{50}$) mg./kg. | $E \times LD_{50}$ |
|---|---|---|---|---|
| Endrin . . . . | 0.56 | $9 \times 10^{-17}$ | 40 | 22.4 |
| Aldrin . . . . | 0.56 | $9 \times 10^{-17}$ | 67 | 37.5 |
| Dieldrin . . . . | 0.48 | $1 \times 10^{-16}$ | 87 | 41.8 |
| Heptachlor . . | 0.40 | $1.2 \times 10^{-16}$ | 90 | 36.0 |
| Lindane . . . . | 0.26 | $2 \times 10^{-16}$ | 125 | 32.5 |
| DDT . . . . . | 0.12 | $4 \times 10^{-16}$ | 250 | 30.0 |
| TDE . . . . . | 0.01 | $5 \times 10^{-15}$ | 3400 | 34.0 |
| Methoxychlor . | 0.005 | $1 \times 10^{-14}$ | 6000 | 30.0 |

[a] Detector temperature: $210°$ C.; carrier gas: nitrogen, 150 ml./min.; detector saturation current: $7 \times 10^{-9}$ ampere.

sensitivities for a number of common chlorinated pesticides; it is evident that, for some of these, the limit of detection with the type of detector available at present, has virtually been attained. It must be emphasized that this discussion is concerned only with detector sensitivity and does not take into account effects due to the column. The extent to which sample is diluted in passage through the column is dependent largely upon column efficiency and retention time. These effects must be taken into account before a practical analytical sensitivity can be determined.

In view of LOVELOCK's work (1962, 1963) with regard to the biological activity of electron absorbing compounds, it is of interest that the acute oral toxicities of the compounds listed parallel quite closely the extent to which they capture electrons. Agreement is remarkably good, considering the impurity of the pesticides, and the uncertainties in the values of acute oral toxicity. It is not to be expected that this correspondence can be extended too far since many other factors, such as transport to the site of action and specific detoxification mechanisms, determine the toxicity of a particular compound.

### f) Linearity of response

Over approximately 40 percent of the current range of the detector response at optimum sensitivity is related to sample weight as follows:

$$I = I_0 \, e^{-kc}$$

where    $I$ = current flowing in the detector at sample peak maximum
        $I_0$ = detector standing current
        $c$ = sample weight
        $k$ = constant

As LOVELOCK (1963) has pointed out, the relationship bears a formal analogy to Beer's Law or to the concentration relationship for a second order reaction.

The linear range and form of the response curve are the same for d.c. and pulsed operation. For a typical detector having a noise level of $3 \times 10^{-12}$ ampere and a saturation current of $8 \times 10^{-9}$ ampere the dynamic range is somewhat less than 1000 : 1. Over the lower 20 percent of the detector range, deviation from linearity is so slight that it can be ignored for all practical purposes and straight line calibration curves may be used.

Fig. 5 shows response curves for d.c. and pulsed operation at various values of applied potential and pulse repetition rate respectively. It will be noted that, although variation in pulse repetition rate has no effect on linearity but merely changes sensitivity, it is not possible to vary the sensitivity of the d.c. polarized detector greatly from the optimum value. At voltages much below optimum, early saturation of the detector occurs with a concomitant negative deviation from linearity. At applied voltages higher than optimum, $S$-shaped response curves are obtained, the deviation from linearity probably being caused by the formation of a negative ion space charge in the detector.

### g) Effect of column bleed

For many of the column materials commonly used, stationary phase or its breakdown products bleeding into the detector cause a reduction in standing current. That the effect is noticeable in both d.c. and pulse

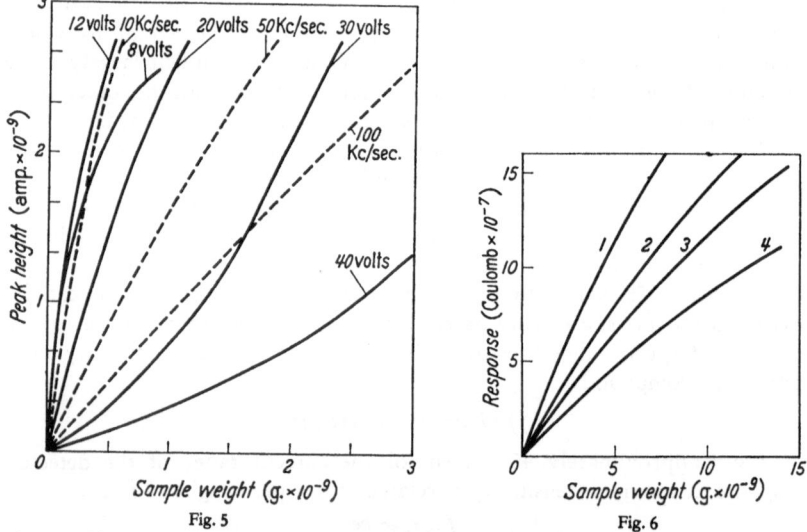

Fig. 5

Fig. 6

Fig. 5. Calibration curves for d.c. and pulsed operation. Detector temperature: 200° C., carrier gas: argon-methane 120 ml./min., and sample: lindane

Fig. 6. Effect of column bleed on sensitivity. Column: 10 percent DEGS on 80 to 90 mesh Anakrom A, 110° C.; detector: 110° C., 150 ml./min. flow rate; sample: diethyl maleate. *1.* Argon-methane carrier; pulse & d.c. Conditioned column. *2.* Argon-methane carrier; pulse & d.c. Unconditioned column. *3.* Nitrogen carrier; d.c. Conditioned column. *4.* Nitrogen carrier; d.c. Unconditioned column

system is evidence that electron capture is responsible for the decrease rather than a change in electron energies as a result of the change in gas composition. The net result is a loss in sensitivity as illustrated in Fig. 6.

In addition, excessive bleed over a prolonged period of time will lead to detector contamination, usually resulting in premature failure of the radioactive source. A similar effect is obtained from the slow elution from the column of sample decomposition products. It cannot be emphasized too strongly that only thoroughly conditioned, stable columns are suitable for use with electron absorption detectors.

When first brought into operation, the standing current-voltage curve for a chosen flow-rate and temperature should be recorded with clean carrier gas only flowing through the detector. Thereafter, comparison with the original curve will immediately indicate deterioration in performance as a result of contamination of the gas stream.

### h) Nature of the carrier gas

Argon containing 10 percent methane is the most suitable of the carrier gases so far studied for operation in the pulsed mode; nitrogen is commonly used for d.c. operation.

For the d.c. mode, sensitivity is greater by a factor of approximately 1.3 when using argon-methane. The difference can be accounted for by the fact that the saturation current for a given detector is higher by a similar amount when the mixed gas is employed. There is comparatively little difference in sensitivity between d.c. and pulsed systems when both employ argon-methane.

Electron mobility is higher in the argon-methane mixture than in nitrogen and, consequently, lower polarization voltages are required for the former gas. Flow and temperature effects connected with ionic mobility are qualitatively the same for both gases; however, the magnitudes of the effects are greater for nitrogen.

### i) Comparison of d.c. and pulse systems

From a theoretical point of view (LOVELOCK 1963), there can be little doubt that electron absorption measurements are best made using thermal electrons under zero field conditions. At the present stage of development of instrumentation, this is most easily done by applying short pulses to the detector, thereby to sample the electron population in the ionization chamber. From the practical standpoint, it is necessary to inquire into the relative merits of pulse and d.c. systems in order to determine whether or not the extra cost and complexity of the pulse system is worthwhile.

The sensitivity of the detector is approximately the same for both systems. What difference exists can be attributed to the nature of the carrier gas used; nitrogen is generally employed with d.c. polarized detectors whereas argon-methane is the preferred carrier gas for pulsed operation. If the highest possible sensitivity is required from a d.c. detector, there is no reason why argon-methane should not be used although it is more difficult to obtain and is more expensive than nitrogen.

Insensitivity to changes in carrier gas flow-rate represents a real advantage of pulsed operation. If, however, a d.c. detector is operated in conjunction with a suitable flow controller and at a reasonably high flow-rate, errors from this source are minimal.

Adjustment for optimum linearity is more critical in the d.c. system and spurious results are obtained if the applied voltage is too high. The ability to change sensitivity by a factor of approximately 10 is an advantage of the pulsed system; this can be done during an analysis, although re-adjustment of the baseline will then be necessary. In many circumstances dilution of the sample with a transparent solvent will serve the same purpose.

Insufficient work has yet been done on temperature effects to assess the relative merits of the two systems. It appears that, for the chlorinated pesticides at least, the pulsed system is less temperature dependent. If adequate temperature control of the detector is available and it is operated at a fixed temperature, little trouble should be experienced from temperature effects.

LOVELOCK (1963) has recently published an authoritative review of the operation of the electron absorption detector and describes in detail possible sources of error associated with d.c. operation of the detector. In practice, it has been found possible to operate the detector over periods of many months without trouble from any of these effects. It must be emphasized, however, that great care must be exercised in the operation of the d.c. system and that from the standpoint of ease of operation, the pulse technique is to be preferred. Otherwise, d.c. operation is perfectly adequate for use in the determination of pesticide residues.

Table II. *Common and chemical names of pesticides mentioned in text*

| | |
|---|---|
| Aldrin . . . . . . | 1,2,3,4,10,10-hexachloro-1,4,4a,5,8,8a-hexahydro-1,4-*endo,exo*-5,8-dimethanonaphthalene |
| DDT. . . . . . . | 1,1,1-trichloro-2,2-bis(*p*-chlorophenyl)ethane |
| Dieldrin . . . . . | 1,2,3,4,10,10-hexachloro-6,7-epoxy-1,4,4a,5,6,7,8,8a-octahydro-1,4-*endo,exo*-5,8-dimethanonaphthalene |
| Endrin . . . . . . | 1,2,3,4,10,10-hexachloro-6,7-epoxy-1,4,4a,5,6,7,8,8a-octahydro-1,4-*endo,endo*-5,8-dimethanonaphthalene |
| Heptachlor . . . . | 1,4,5,6,7,8,8-heptachloro-3a,4,5,5a-tetrahydro-4,7-*endo*-methanoindene |
| Lindane . . . . . | Gamma isomer of 1,2,3,4,5,6-hexachlorocyclohexane |
| Methoxychlor . . . | 1,1,1-trichloro-2,2-bis(*p*-methoxyphenyl)ethane |
| TDE (DDD) . . . | 1,1-dichloro-2,2-bis(*p*-chlorophenyl)ethane |

## Summary

The introduction of the electron absorption detector represented a major advance in gas chromatographic detection systems. The detector is not only the most sensitive device available for the detection of certain types of compounds but exhibits a high degree of selectivity. As with most ionization detectors, the mechanics of operation of the device are complex and, as yet, but imperfectly understood. Empirical study of the effect of experimental variables upon detector performance has now yielded sufficient information to allow the confident use of the device for the quantitative determination of pesticide residues. The dependence of detector response upon applied potential, nature and flow rate of carrier gas and temperature, the linear range and sensitivity have been determined, and pulse and d.c. modes of operation are compared.

## Résumé *

L'introduction du détecteur à capture d'électrons a déterminé un grand progrès dans les systèmes de détection de la chromatographie en phase gazeuse. Ce détecteur est non seulement celui qui est actuellement le plus sensible pour la détection de certains types de composés, mais encore il se montre hautement sélectif. Comme avec la plupart des détecteurs à ionisation son fonctionnement est complexe et, pour l'instant, imparfaitement compris. L'étude empirique de l'effet des variables expérimentales sur les performances du détecteur a maintenant donné des informations suffisantes pour que l'analyse quantitative des résidus de pesticides puisse être confiée à ce dispositif. Les relations entre la réponse du détecteur et le potentiel appliqué, la nature et le débit du gaz vecteur et la température ainsi que la sensibilité et l'étendue du domaine de réponse linéaire ont été précisées. Les fonctionnements en continu et avec pulsations ont été comparés.

## Zusammenfassung **

Die Einführung des Elektronen-Absorptions-Detektors bedeutet einen erheblichen Fortschritt auf dem Gebiet der Detektoren für die Gaschromatographie. Dieser Detektor ist nicht nur der z. Z. empfindlichste zum Nachweis bestimmter Verbindungstypen, sondern er arbeitet auch mit einem Höchstmaß an Selektivität. Wie bei den meisten Ionisations-Detektoren ist die Wirkungsweise des Instruments komplex und noch nicht völlig aufgeklärt. Empirische Untersuchungen über den Einfluß variierter Versuchsbedingungen auf die Wirkungsweise des Detektors haben aber inzwischen genügend Aufklärung gebracht, um eine zuverlässige Verwendung des Gerätes für quantitative Analysen von Pesticid-Rückständen zu gestatten. Es wurden bestimmt: Abhängigkeit der Detektorreaktion von der angelegten Spannung, Art und Fließgeschwindigkeit des Trägergases und der Temperatur sowie der Linearitätsbereich und die Empfindlichkeit; außerdem wurden die Arbeitsweisen unter pulsierendem und normalem Gleichstrom miteinander verglichen.

## References

BELLAR, T. A., and J. E. SIGSBY, JR.: Application of electron capture detection to gas chromatography in air pollution. 144th National Meeting, Amer. Chem. Soc. Los Angeles. April 1963.

CLARK, S. J.: Gas chromatographic analysis of pesticide residues using the electron affinity detector. 140th National Meeting, Amer. Chem. Soc., Chicago. September 1961.

COULSON, D. M.: Electron capture detection of pesticides. S. R. I. Pesticide Research Bull. 2, 1 (1962).

DARLEY, E. F., K. A. KETTNER, and E. R. STEPHENS: Analysis of peroxyacyl nitrates by gas chromatography with electron capture detection. Anal. Chem. 35, 589 (1963).

DAWSON, H. J.: Determination of methyl-ethyl lead alkyls in gasoline by gas chromatography with an electron capture detector. Anal. Chem. 35, 542 (1963).

---

* Traduit par R. MESTRES.
** Übersetzt von H. FREHSE.

GOODWIN, E. S., R. GOULDEN, and J. G. REYNOLDS: Rapid identification and determination of residues of chlorinated pesticides in crops by gas-liquid chromatography. Analyst 86, 697 (1961).

LANDOWNE, R. A., and S. R. LIPSKY: Electron capture spectrometry, an adjunct to gas chromatography. Anal. Chem. 34, 726 (1962).

LOVELOCK, J. E.: Affinity of organic compounds for free electrons with thermal energy: its possible significance in biology. Nature 189, 729 (1961).

— Ionization methods for the analysis of gases and vapors. Anal. Chem. 33, 162 (1961).

— Electron absorption detectors and technique for use in quantitative and qualitative analysis by gas chromatography. Anal. Chem. 35, 474 (1963).

—, and S. R. LIPSKY: Electron affinity spectroscopy. — A new method for the identification of functional groups in chemical compounds separated by gas chromatography. J. Amer. Chem. Soc. 82, 431 (1960).

—, P. G. SIMMONDS, and W. J. A. VANDENHEUVEL: Affinity of steroids for electrons with thermal energies. Nature 197, 249 (1963).

—, and A. ZLATKIS: A new approach to lead alkyl analysis: Gas phase electron absorption for selective detection. Anal. Chem. 33, 1958 (1961).

— —, and R. S. BECKER: Affinity of polycyclic aromatic hydrocarbons for electrons with thermal energies: Its possible significance in carcinogenesis. Nature 193, 540 (1962).

WATTS, J. O., and A. K. KLEIN: Determination of chlorinated pesticide residues by electron capture gas chromatography. J. Assoc. Official Agr. Chemists 45, 102 (1962).

WENTWORTH, W. E., and R. S. BECKER: Potential method for the determination of electron affinities of molecules: Application to some aromatic hydrocarbons. J. Amer. Chem. Soc. 84, 4263 (1962).

# Selective detection and identification of pesticide residues

By

Theron Johns * and Charles H. Braithwaite, jr. **

With 7 figures

## Contents

## I. Introduction

Since the introduction in the early 1940's of organic chemicals for pest control the amounts and types of chemicals used for this purpose have increased rapidly. The use of these chemicals has permitted the production of higher quality agricultural products with greater yields per acre. However, these advantages have not been achieved without corresponding disadvantages. Organic chemicals are more difficult to remove from food products than the inorganic water soluble salts previously used. This has created real potential residue problems which have, in turn, caused the enactment of stringent regulations regarding the permissible concentration levels of specific pesticides in food products.

Various analytical methods have been proposed and used in order to insure that food products meet government regulations in regard to pesticide residues. Since chlorinated hydrocarbons represent the major type of organic pesticides, many of the methods developed are more or less specific for this type of chemical. Some methods can be used satisfactorily for the measurement of specific compounds while others measure the total amount of possibly several combined pesticides through determination of organically-bound chlorine. The well-known SHECHTER-HALLER colorimetric test is an example of methods of this type. Other methods of this type include the oxygen flask method described by St. John and Lisk (1961), the combustion-titration method described by Gunther and Blinn (1955), and the neutron activation method described by Guinn and Wagner (1960).

---

* Beckman Instruments, Inc., Fullerton, California.
** Cal-Colonial Chemical Company, Orange, California.

Recently additional methods have been developed which are capable of measuring each of the pesticides individually. Of these, the paper chromatography method described by MILLS (1959) is probably the most widely used. It has also been shown that gas chromatography is applicable to the analysis of mixtures of pesticides or pesticide residues. Potentially gas chromatography has the advantage of better separation, higher sensitivity, and more accurate quantitation.

The first use of gas chromatography for the analysis of pesticides was reported by ADLARD and WHITHAM (1958). A thermal conductivity detector was used in this investigation. This type of detector is adequate for the analysis of mixtures of pesticides. However, the sensitivity is too limited for work with pesticide residues. Additionally, since the detector is non-selective, an unreasonable amount of clean-up is required to prevent interference from naturally occurring compounds. COULSON et al. (1960) recognized these limitations and developed a selective, sensitive, micro-coulometric detector which made it possible to analyze pesticide residues with minimal sample cleanup. Even more sensitive selective detection has been achieved with a modified LOVELOCK argon ionization detector. For example GOODWIN et al. (1960) found that, if they operated this detector at a low applied potential, chlorinated pesticides could be detected at about the 0.01 part per million (p.p.m.) level without significant interference from naturally occurring compounds.

It is also advantageous to combine sensitive non-selective detection with sensitive selective detection for more complete qualitative and quantitative information on the sample. Occasionally some doubt might still exist in regard to the identity of component or components measured. It may also be desirable to establish the identity of breakdown or metabolic products. In these instances an acceptable reliable qualitative method is required. Two techniques were evaluated and are herein described which meet these requirements.

## II. Experimental

**Apparatus.** All gas chromatographic measurements were made with a Beckman Model GC-2 Gas Chromatograph equipped with a combination flame ionization-solution conductivity detector. Infrared spectra were obtained with a Beckman IR-9 Spectrophotometer equipped with a micro-specular reflectance attachment.

**Samples and reagents.** Samples were either synthetic preparations or residues from milk or butterfat. The reagents used were research grade or spectro-quality.

**Operating conditions of the gas chromatograph.** All analytical runs were made using a six-foot $1/8$-inch aluminium column packed with six percent S.E. 30 silicone plus 0.5 percent Epon 1001 on hydrochloric acid-washed 42 to 48 mesh Chromosorb W. The column was operated at 190° C. and at a flow rate of 50 ml./min. of helium. Preparative runs were made with a six-foot $1/4$-inch aluminum column packed with 10 percent S.E. 30 silicone plus 0.5 percent Epon 1001 on hydrochloric acid-washed 42 to 48 mesh Chromosorb W. The column was operated at 220° C. with o flow rate of about 200 ml./min. of helium. The effluent was split at the column exit at a ratio of 20 parts to the collector and one part to the detector.

### III. Discussion and results

A schematic diagram of the combination hydrogen flame-electrical or solution conductivity detector is shown in Fig. 1. The hydrogen flame burner serves as a combustion chamber to convert the organic pesticides into

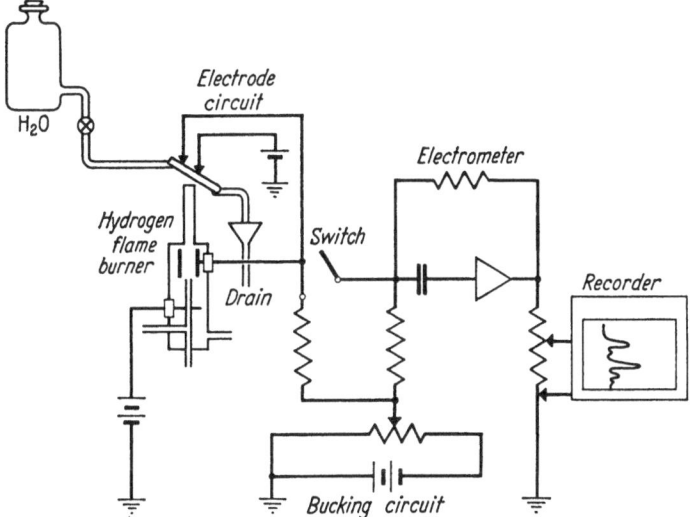

Fig. 1. Schematic of electrical conductivity detector

gaseous products. These combustion gases are exhausted through a chimney so that a percentage of the gases comes in contact with a flowing film of conductivity solution-usually distilled water. If combustion products are formed which are soluble and ionize in the conductivity solution, a change in electrical conductivity of the solution results. If a constant potential is applied between two electrodes in contact with the solution and current flow is measured, a change in current will be observed when a change in the conductivity of the solution occurs. The essential features of the detector then are:

1. A conductivity solution with suitable background characteristics.
2. A means of exposing a flowing film with favorable dimensions, shape, and velocity to the exhaust gases.
3. A combustion chamber.
4. Electrodes of proper size and spacing.
5. An applied potential between the electrodes.
6. A means of measuring the current change which results from a change in the conductivity of the solution. Under the conditions used the background current is low, in the range of $1-5\times10^{-8}$ ampere. Therefore, an electrometer is used for measuring the current.

The trolley system used for exposing the film of conductivity solution to the combustion gases is shown in Fig. 2. Spacing between the trolley bars is approximately 0.5 mm. The electrodes are placed about 1 cm. apart.

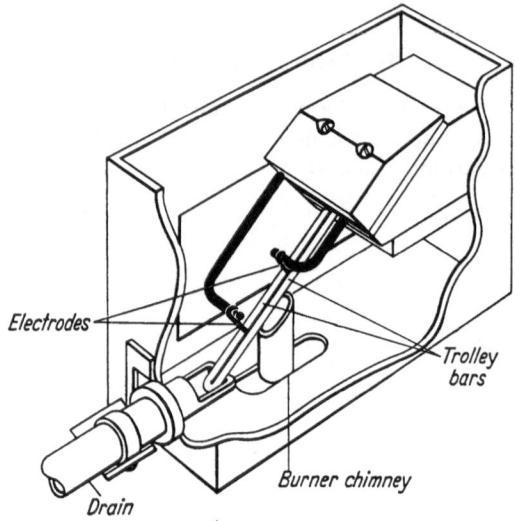

Electrodes

Trolley bars

Burner chimney

Drain

Fig. 2. Schematic of detector box

A typical flow rate of the conductivity solution is 0.25 ml./min. Some of the factors which affect sensitivity are:

1. Film width and thickness.
2. Flow rate of the conductivity solution.
3. Spacing of the electrodes.
4. Magnitude of applied potential.
5. Nature of the exhaust gases.

Factors which affect noise and background stability are:

1. Variations in velocity of the conductivity solution.
2. Temperature fluctuations.
3. Conductivity of the solution.

A comparison of the results which can be obtained with selective electrical or solution conductivity detector and the non-selective flame ionization detector are shown in Fig. 3. In each case a sample of 0.5 $\mu$l. of hexane containing 0.5 $\mu$g. of each of five pesticides was used. It can be seen that the signal for hexane relative to the signal for the pesticides is much lower with the conductivity detector than with the flame detector. Some signal is obtained with hexane with the conductivity detector since one of the combustion products is carbon dioxide. Carbon dioxide has relatively limited solubility in water and the rate of hydrolysis or ion formation is slow. Therefore, the sensitivity for a compound which contains only carbon, hydrogen, and oxygen is very low compared to a compound which contains halogens or sulfur. In this case the sensitivity for lindane ($\gamma$-1,2,3,4,5,6-hexachlorocyclohexane) is about four-thousand times greater than the sensitivity for hexane. It is also apparent that a greater attenuation is required to keep the peaks on scale for the solution detector than for the flame detector. This could be interpreted to mean that the sensitivity of the

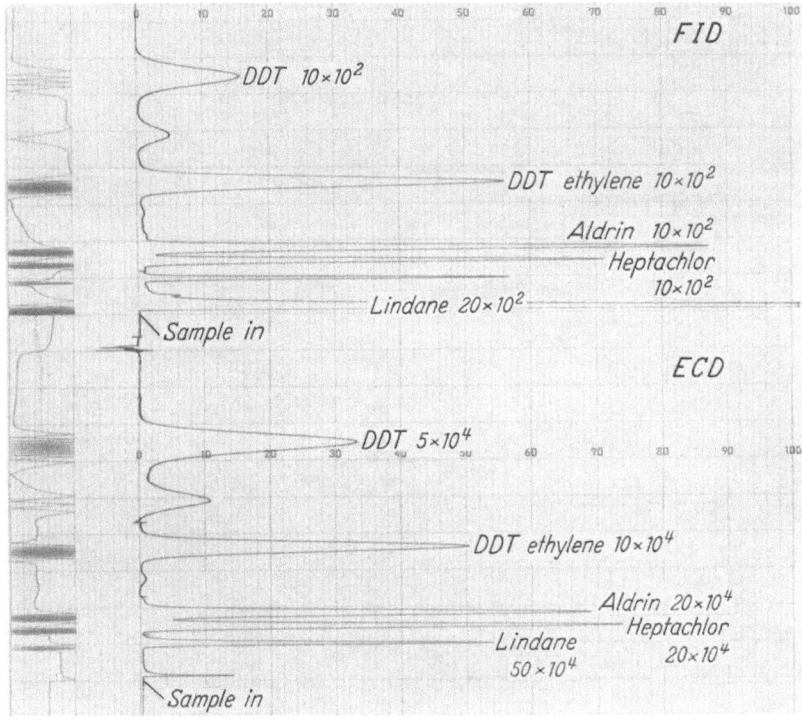

Fig. 3. Comparison of sensitivity and resolution of flame ionization (*FID*) and electrical conductivity (*ECD*) detectors. Each sample was 0.5 μl. containing 0.5 μg. of each component

solution conductivity detector is greater. However, the only meaningful quantity in evaluating the sensitivity of a detector is signal-to-noise ratio. As used, the conductivity detector had a background current of about $5 \times 10^{-8}$ ampere. The electrometer used had a sensitivity of $5 \times 10^{-13}$ ampere full scale at $X1$ attenuation. This means that, if a noise level of only one percent is to be tolerated and one wanted to use the detector at $X1$ attenuation, the background current stability would have to be $5 \times 10^{-15}$ ampere. With a background of $5 \times 10^{-8}$ ampere this would require a stability of one part in $10^7$. A stability of this magnitude would be extremely difficult to achieve with any high impedance system and, in this case, with the simple design of the detector used the stability is about one part in $10^3$. This permits operation down to an attenuation of $1 \times 10^4$ with a noise level of about one percent.

The sensitivity of the detector increases as the flow rate of the conductivity solution is decreased. However, if the film velocity is decreased too much a loss of resolution between closely spaced peaks will occur. In this case a flow of 0.25 ml./min. was found to be about the minimum which could be used without a significant loss of resolution between heptachlor

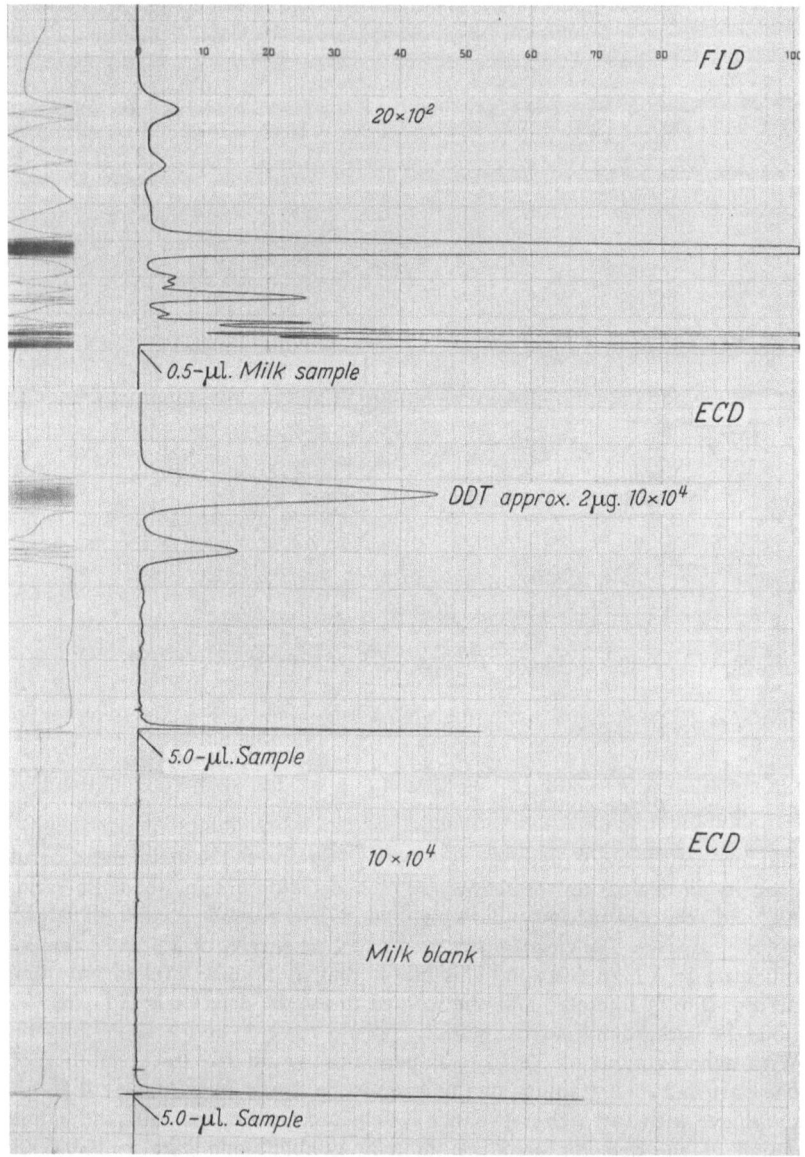

Fig. 4. DDT from milk, flame ionization (*FID*) and electrical conductivity (*ECD*) detectors. 50 μg. of DDT was added to 10 ml. of milk, and the extracted residue was taken up in 100 μl. of hexane

(1,4,5,6,7,8,8-heptachloro-3a,4,7,7a-tetrahydro-4,7-methanoindene) and aldrin (1,2,3,4,10,10-hexachloro-1,4,4a,5,8,8a-hexahydro-1,4-*endo,exo*-5,8-

dimethano-naphthalene). The detection limit for DDT [2,2-bis($p$-chloro-phenyl)-1,1,1-trichloroethane] with the solution conductivity detector as used with a signal twice noise was $6 \times 10^{-9}$ g. This corresponds to a detection limit of about $2 \times 10^{-10}$ g. of chlorine per second.

The selectivity of the solution conductivity detector was evaluated with an extract of DDT which had been added to milk. The results of this test are shown in Fig. 4. A 10-ml. aliquot of milk was extracted with three 20-ml. fractions of hexane. The hexane fractions were then combined and evaporated nearly to dryness. The residue was taken up in 100 $\mu$l. of hexane and a 5.0-$\mu$l. aliquot was injected into the gas chromatograph. This was the blank for the solution conductivity detector and the only peak was a small signal for the hexane. This procedure was repeated with a 10-ml. aliquot of milk to which 50 $\mu$g. of DDT had been added. Aliquots of the residue were chromatographed with both the solution conductivity detector and the flame ionization detector. With the solution conductivity detector significant peaks were obtained only for hexane, DDT, and an impurity which was initially present in the DDT. The flame ionization showed many other components. Some of these components are believed to have been in the pure hexane used and were concentrated by the sample preparation procedure. Another possibility, since the milk was taken from a waxed carton, is that these components may have come from the wax. A later sample of milk which was taken from a glass bottle and extracted with chloroform did not show these components.

Repeatability with the solution conductivity detector is very good. This, coupled with high selectivity, makes it a good quantitative detector for pesticides. It is necessary, however, to calibrate under fixed operatino conditions. A typical response curve is shown in Fig. 5 for varying amounts of

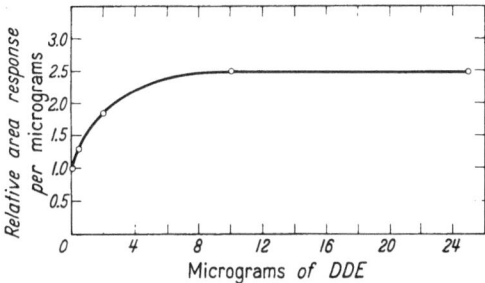

Fig. 5. Response curve for DDT-ethylene (DDE) with electrical conductivity detector

DDT-ethylene [2,2-bis($p$-chlorophenyl)-1,1-dichloroethylene]. This method of plotting response is unusual since it exaggerates any non-linearity in detector response. It shows that the response of the detector per unit amount of component is decreased as the amount of component is decreased. As the amount is ancreased the response per unit amount approaches a limiting maximum value and linearity is good. Therefore, for quantitation of small amounts of pesticides the detector should be calibrated over the concentration range of interest. It should also be remembered that this response curve was obtained by putting known quantities of DDT-ethylene in the gas chromatograph. It reflects any non-linearity due to breakdown or hold-up of DDT-ethylene in the inlet or column in addition to the characteristics of the detector. The data for the response curve for DDT-ethylene are given in Table I.

Table I. *Calibration data for DDT-ethylene*

| Micrograms DDE | Area units | Area units per microgram | Relative response per microgram |
|---|---|---|---|
| 0.04 | 700 | 1750 | 1.00 |
| 0.4 | 900 | 2250 | 1.29 |
| 2.0 | 6520 | 3260 | 1.86 |
| 10.0 | 43600 | 4360 | 2.49 |
| 25.0 | 109000 | 4360 | 2.49 |

The gas chromatographic-solution conductivity method of analysis was utilized to evaluate several samples of butterfat from different sources over an extended period of time. A simplified Florisil cleanup method was used to reduce the amount of butterfat injected into the column. Approximately 0.5 g. of butterfat was weighed to the nearest milligram and then dissolved in 25 ml. of hexane. The hexane solution was then passed into a 12-inch column of Florisil containing 30 pecent by weight of anhydrous sodium sulfate. The Florisil column was developed with 150 ml. of a hexane-diethyl ether solution containing six percent diethyl ether. The volume of this solution was reduced to 25 $\mu$l. and the entire amount injected into the gas chromatograph. The results on samples from several dairies are given in Table II.

Table II. *Pesticide residue levels of typical butterfat samples*

| Dairy | March 8, 1963 | | March 18, 1963 | |
|---|---|---|---|---|
| | DDT (p.p.m.) | Total (p.p.m.) | DDT (p.p.m.) | Total (p.p.m.) |
| A | 0.4 | 0.7 | 0.1 | 0.2 |
| B | 0.4 | 0.8 | 0.1 | 0.1 |
| C | 0.5 | 1.5 | 0.3 | 0.5 |
| D | 0.9 | 1.8 | 0.2 | 0.5 |
| E | 0.4 | 0.7 | 0.4 | 0.8 |
| F | 0.1 | 0.2 | 0.2 | 0.4 |
| G | 0.4 | 0.9 | 0.2 | 0.9 |
| H | 0.1 | 0.3 | 0.1 | 0.2 |

The results in Table II indicate that, when residue levels are determined and found too high, production procedures can then be adopted which reduce the amount of pesticides in milk or butterfat. Dairy *B* had a history of high residue levels and corrective measures had been taken just prior to the first tests. Dairy *D* had put 11 new cows of unknown pesticide content on the milking line a few weeks before the first test.

The problem of a positive identification of pesticide residue components was also considered in this investigation. If the individual properties of retention time, selective detection, and the ratio of response of a selective detector to the response of a non-selective detector are considered together, a reasonable assumption on identification is possible by gas chromatography. However, in some instances further proof may be desirable. Infrared spectrophotometry is capable of giving a positive identification and it has been used for this purpose in the analysis of pesticide residues. In most instances this technique has been inconvenient for residue analyses because of the relatively large sample sizes required.

In the present work this difficulty was reduced using a new infrared technique recently described by SLOANE *et al.* (1963). The sample is collected in a suitable solvent and then transferred to a reflecting mirror. The solvent containing the pesticide is injected into a small spot on the mirror with a micro-syringe. The solvent is evaporated as the sample is added to the mirror. The mirror spot is then examined using a microspecular reflectance attachment. With this technique the light makes a double pass through the sample and, therefore, greater sensitivity is obtained than would be the case with micro-pellets or with solutions. The infrared spectra of four pesticides at the 50-$\mu$g. level each are shown in Fig. 6. This method was also

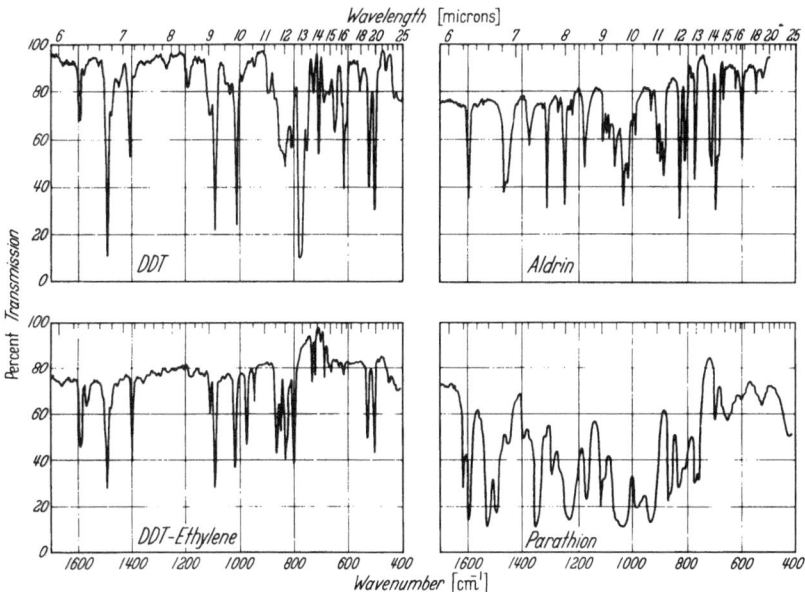

Fig. 6. Infrared spectra of 50 $\mu$g. quantities of four compounds by micro-specular reflectance

used to examine DDT which had been extracted from milk. The results are shown in Fig. 7. An aliquot of the extract was first examined directly by infrared ($A$). Since no cleanup was used, the infrared spectrum was chiefly that of triglycerides from milk. When a second aliquot equivalent to about 50 $\mu$g. of DDT was put through the gas chromatograph ($B$) the infrared spectra of the collected fraction ($C$) was essentially that of pure DDT.

In this case the gas chromatograph served as the cleanup technique for the sample. However, this and subsequent samples established that injecting significant amounts of butterfat into the gas chromatograph should be

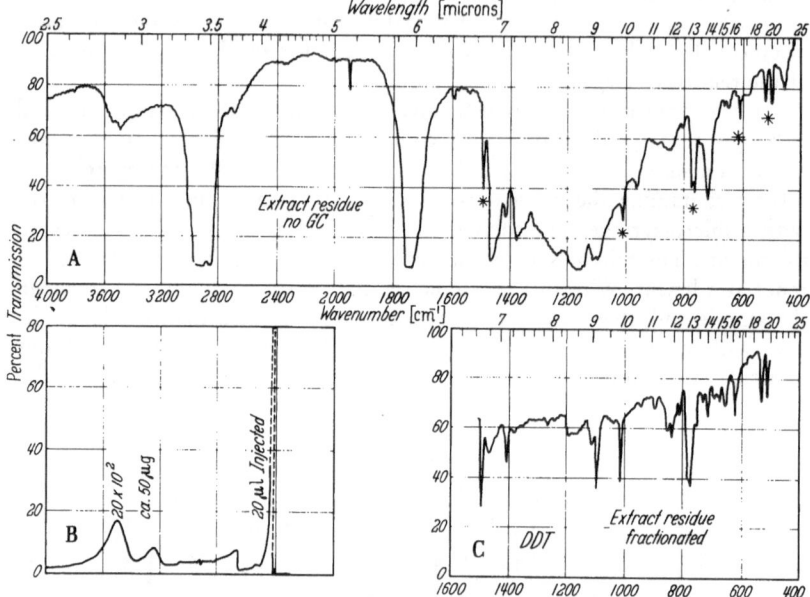

Fig. 7. Infrared spectra of DDT from milk (butterfat)

avoided. The butterfat prevents a clean injection of the pesticides into the chromatographic column and after several samples the characteristics of the column are modified. This results in increased retention times, peak tailing, and decreased resolution. It appears then that even with selective detection some sample cleanup is required.

This requirement might be avoided by using an extraction technique which is favorable for gas chromatography. The authors suggest that one method which might be worthwhile is an acid hydrolysis of the triglycerides. This would yield glycerol and fatty acids which should not interfere in the determination of the pesticides.

## Summary

The solution conductivity detector in combination with a hydrogen flame detector is a useful selective detector for the gas chromatographic determination of pesticide residues. This detector measures the change in conductivity of a solution caused by exposing a film of the solution to the combustion products of the sample components. As described, the detector has a detection limit of about $10^{-10}$ gram per second of sulfur or halogens. This detection system in combination with gas chromatography gave reasonable assurance as to the identity of the components measured. However, more positive proof was obtained with a new infrared technique using a micro-specular attachment which reduced the amount of sample needed to the 25 to 50 microgram level.

## Résumé *

La combinaison d'un détecteur mesurant la conductivité d'une solution à un détecteur à ionisation de flamme d'hydrogène est intéressante pour la détermination sélective des résidus de pesticides par chromatographie en phase gazeuse. Ce détecteur mesure les variations de conductivité d'un film de solution exposé aux produits de combustion des composants de l'échantillon. La sensibilité limite de ce détecteur est d'environ $10^{-10}$ g de soufre ou d'halogène par s. Ce système de détection, en combinaison avec la chromatographie en phase gazeuse elle même, donne une assurance raisonnable de l'identité des composés analysés. Cependant, une nouvelle technique de spectrophotométrie infra-rouge utilisant un équipement qui permet de réduire à 25—50 microgrammes la quantité nécessaire d'échantillon donne une preuve plus positive encore.

## Zusammenfassung **

Der Leitfähigkeitsdetektor, in Kombination mit einem Flammenionisations-Detektor, ist wertvoll als selektiver Detektor zur gaschromatographischen Bestimmung von Pesticid-Rückständen. Er mißt die Leitfähigkeitsänderung in einer Lösung, die sich ergibt, wenn ein dünner Film der Lösung den Verbrennungsprodukten der zu analysierenden Probe ausgesetzt wird. Wie beschrieben, hat der Detektor eine Nachweisgrenze von $10^{-10}$ Gramm pro Sekunde für Schwefel oder Halogene. Dieses Detektorsystem lieferte bei der Gaschromatographie hinreichende Bestätigung der Identität der gemessenen Verbindungen. Eine weitergehende Sicherung ließ sich jedoch mittels einer neuen Infrarot-Technik unter Verwendung eines „micro-specular"-Zusatzes erzielen, wodurch die benötigte Substanzmenge auf 25—50 Mikrogramm reduziert wird.

## References

ADLARD, E. R., and B. T. WHITHAM: Applications of high temperature gas-liquid chromatography in the petroleum industry. Gas Chromatography (Ed. D. H. DESTY). London: Butterworths 1958.

COULSON, D. M., L. A. CAVANAGH, J. E. DEVRIES, and B. WALTHER: Microcoulometric gas chromatography of pesticides. J. Agr. Food Chem. 8, 399 (1960).

GOODWIN, E. S., R. GOULDEN, A. RICHARDSON, and J. G. REYNOLDS: The analysis of crop extracts for traces of chlorinated pesticides by gas-liquid partition chromatography. Chem. & Ind. 39, 1220 (1960).

GUINN, V. P., and C. D. WAGNER: Instrumental neutron activation analysis. Anal. Chem. 32, 317 (1960).

GUNTHER, F. A., and R. C. BLINN: Analysis of insecticides and acaricides. New York: Interscience 1955.

MILLS, P. A.: Detection and semiquantitative estimation of chlorinated organic pesticide residues in foods by paper chromatography. J. Assoc. Official Agr. Chemists 42, 734 (1959).

* Traduit par R. MESTRES.
** Übersetzt von H. FREHSE.

SLOANE, H. J., T. JOHNS, W. J. CADMAN, and W. F. ULRICH: Infrared examination
of micro samples from gas chromatographic effluent. Presented 14th Pittsburgh
Conference, Analytical Chemistry and Applied Spectroscopy, March 1963.
ST. JOHN, L. E., JR., and D. J. LISK: Modified and improved procedure for
Schöniger total chlorine residue analysis. J. Agr. Food Chem. 9, 468 (1961).

# Applications of the microcoulometric titrating system as a detector in gas chromatography of pesticide residues*

JACK A. CHALLACOMBE ** and JAMES A. McNULTY **

With 8 figures

## Contents

## I. Introduction

The pesticide residue chemist always has two questions before him-what is it and how much is there? Previous papers have pointed out that gas chromatography is a very valuable tool for pesticide residues. This technique does a very good job of separating complex mixtures. Also pointed out is the need for a specific detector, because the residue chemist is plagued with associated crop extracts that probably never are completely removed from a sample. The electron affinity and electrical conductivity devices are highly discriminatory-type detectors which have been used for pesticide analysis.

The absolute specificity and quantitative nature of microcoulometry are two distinct aids to the pesticide residue chemist. The chemist would like to have a column that would completely resolve from each other the hundreds of compounds suggested for use as crop sprays. As yet such a column has not been described. Those commonly employed elute halogen- and sulfur-containing insecticides without much discretion. Using a detector which will see only one class of these compounds, therefore, eliminates half the problem (CASSIL 1962). Some of the insecticides contain both halogen and sulfur, which is a further aid in identification when microcoulometry is employed with gas chromatography.

---

* Presented in part at the 144th National Meeting of the Amer. Chem. Soc., Div. of Agr. and Food Chem., Los Angeles, California, April, 1963.
** Dohrmann Instruments Company, 990 Varian Street, San Carlos, California.

The Microcoulometric Gas Chromatograph manufactured by Dohrmann Instruments was developed to make possible the simultaneous qualitative and quantitative analysis for pesticide residues on foodstuffs in as little time as possible. During the past few years this system has become well accepted in leading residue laboratories throughout the country. Descriptions of this instrumentation and operational parameters have been presented at many technical meetings and appear in scientific publications (BURKE and JOHNSON 1962, CASSIL 1962, COULSON and CAVANAGH 1960, COULSON et al. 1960, ERICKSON and HIELD 1962), as well as in company literature.

Many residue chemists expressed a desire for a higher degree of sophistication in the gas chromatograph portion of this instrumental system. Among the features desired in the gas chromatograph were multiple column operation, varied and versatile detectors, and flexibility in sample introduction — in other words, a system which would do other jobs as well as pesticide residues.

Inasmuch as there are a multitude of manufacturers of general-purpose gas chromatographs, we elected to develop a system that would allow the chemist to utilize the extremely specific and quantitative features of microcoulometry with the particular chromatograph he feels will fulfill his need. The system should also be a complete analytical device in itself.

## II. Development of the microcoulometric titrating system

The Microcoulometric Titrating System was developed to attach to the chromatograph which would have the desired features. Installations of the Microcoulometric Titrating System have been made on the Beckman Thermotrac, F & M 500, Perkin-Elmer 800, and a number of Micro-Tek 2500's, as well as custom-made chromatographs.

The System consists of the Automatic Chloride Analyzer originally described by COULSON and CAVANAGH (1960) with slight modifications in circuitry, two new improved titration cells — one specific for halogens (except fluorine) and the other for sulfur — and a Model S-100 combined inlet/combustion unit.

The titration cell contains four electrodes. For the sulfur determination, the electrolyte used is 0.04 to 0.06 percent potassium iodide in 0.4 percent acetic acid. Metal electrodes for the sulfur determination are all platinum. The reference electrode is platinum in saturated triiodide. The metal electrodes for the halogen determination are all silver except the generator cathode, which is a platinum spiral. The electrolyte for halogen is 70 percent acetic acid. The reference electrode is silver in saturated silver ion.

In the determination of halogen-and sulfur-containing compounds, the sample is burned in the combustion zone at about 800° C. in an excess of oxygen. The effluent gases from the combustion tube enter the titration cell largely as carbon dioxide, water vapor, hydrogen halide, and sulfur dioxide. Halogen entering the titration cell is precipitated as silver halide. Sulfur dioxide entering the titration cell is oxidized to sulfur trioxide by the triiodide.

Changes in titrant concentration caused by sample entering the titration cell alter the half-cell potential and give an error signal to the high-gain, null-balance servo amplifier in the coulometer. The servo system then

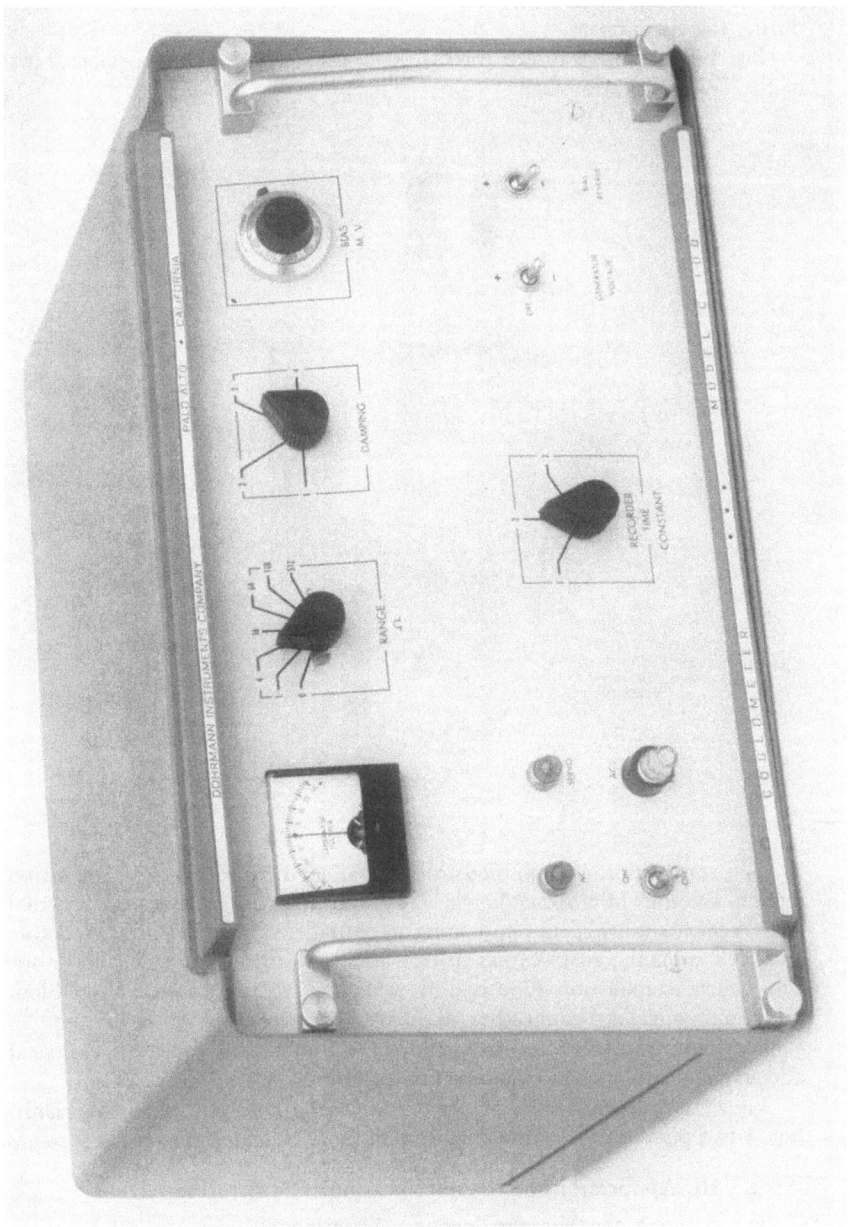

Fig. 1. The Model C-100 Coulometer

applies a voltage to the generator electrodes to replace immediately that titrant which was consumed by the sample. The current passing between the generator electrodes to maintain concentration flows through a precision resistor network. A recorder monitoring the IR drop of this resistor network will then display the total number of coulombs used during the titration.

Fig. 1 is a photograph of the C-100 Coulometer. Fig. 2 is a photograph of the T-200 Titration Cell.

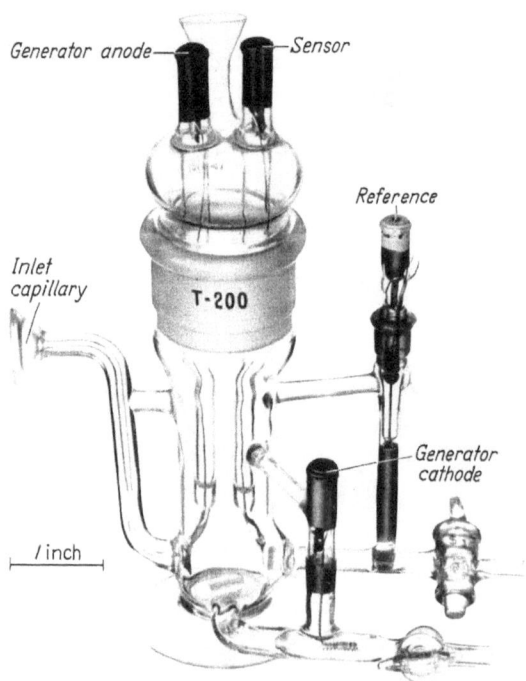

Fig. 2. Model T-200 Titration Cell

The combustion unit employs all the salient features of the combustion section of the Microcoulometric Gas Chromatograph. Major structual improvements allowed the packaging in a single, standard width, stackable unit. All coolant and gas lines utilize double-ferrule Swagelok connectors. The quartz combustion tube is also connected to the volatilization block with a Swagelok connector. Improved furnace design provides an exellent temperature profile throughout the combustion zone. All electrical and mechanical connections are easily accessible.

Fig. 3 is a photograph of the S-100 Sample Inlet/Combustion Unit. Fig. 4 is a photograph of the complete Microcoulometric Titrating System.

### III. Applications of the microcoulometric titrating system

Fig. 5 shows the Microcoulometric Titrating System in our laboratory connected to a Micro-Tek Model 2503 Research chromatograph. This model is a version of the 2500 R Series Gas Chromatograph which is designed for pesticide residue analysis. It incorporates aluminum construction, heat-traced sample transfer lines, and the special removable inlet sleeves as described by CASSIL (1962). These inlets have proven highly successful in residue work as well as in other fields of general gas chromatography.

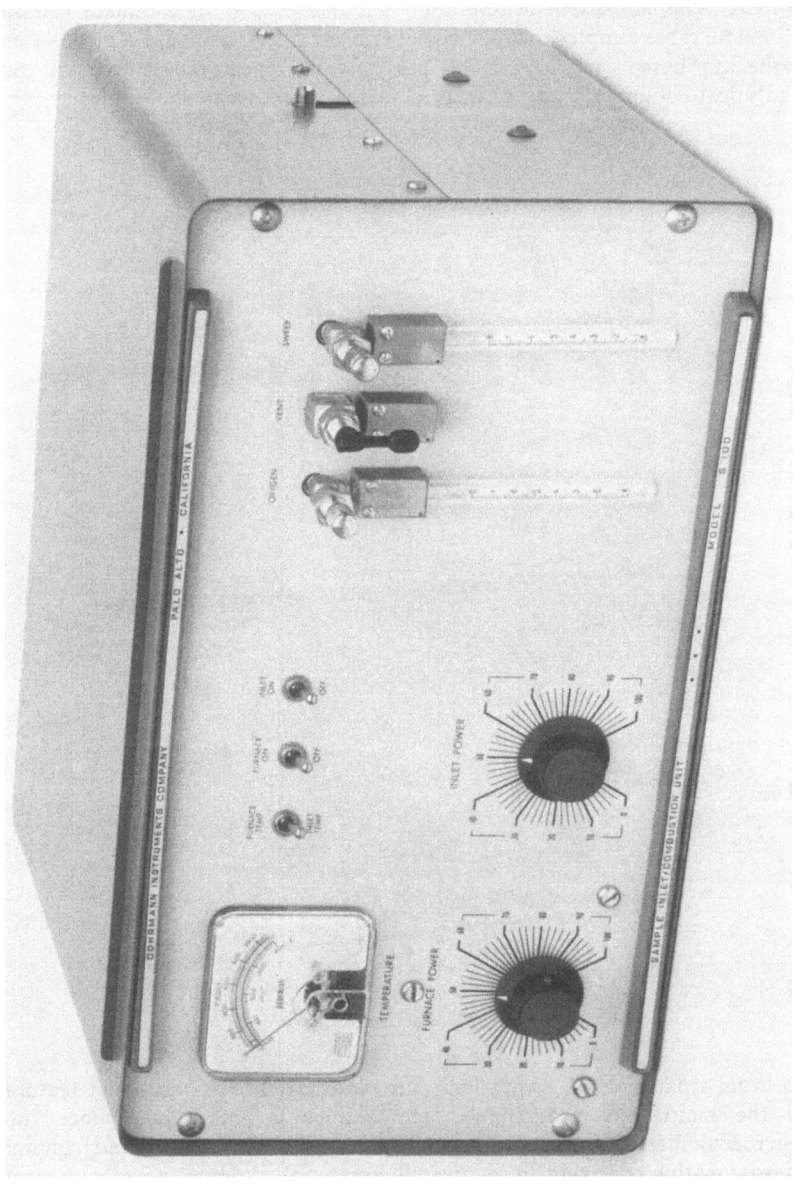

Fig. 3. Model S-100 Sample Inlet / Combustion Unit

A chromatograph of this type is of invaluable aid to the residue chemist who is faced with a daily workload of completely unknown samples. It is a multiple-column instrument, allowing the residue chemist to have ready for immediate analysis different analytical columns to aid in identification. Another approach taken by many users of this equipment is to employ a short column 10-inches or 1-foot long as a rapid-screening column. This affords little resolution but allows a 4- or 5-minute run to eliminate those samples that are not of further interest. Should a peak — or rather a "hump" — appear, he has only to repeat the sample on the analytical column already connected to the second removable inlet system

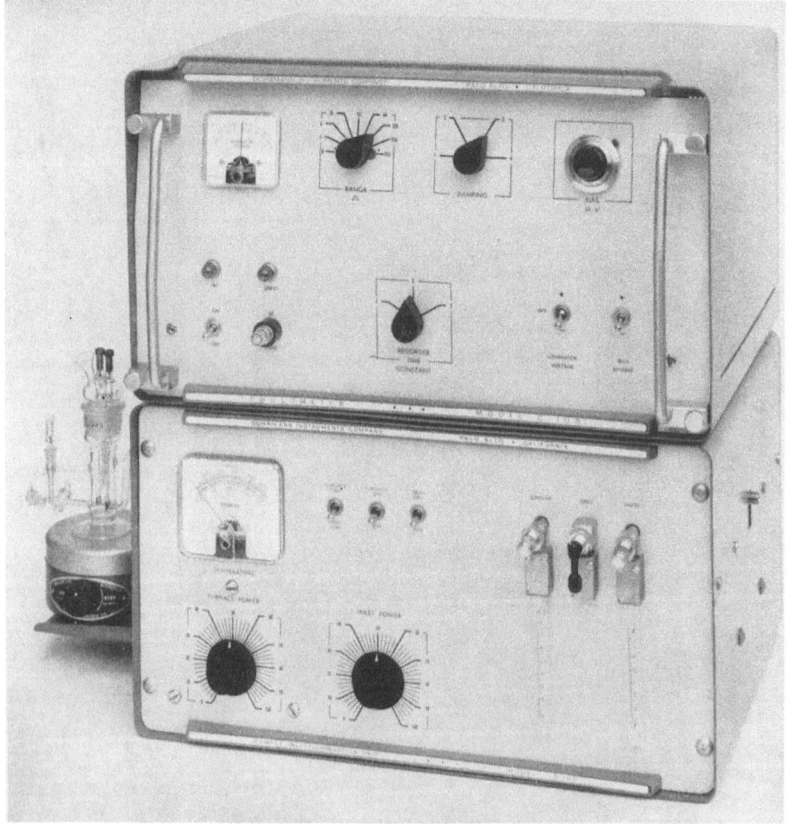

Fig. 4. Complete Microcoulometric Tritrating System

to begin separation, identification, and quantitation. An additional feature is the multilinear programmed temperature column oven. Since the Microcoulometric Titrating System is almost flow-insensitive and column bleed contains relatively little titratable material, a programmed run from 150 to 260° C. will increase the baseline only 1 or 2 percent at half

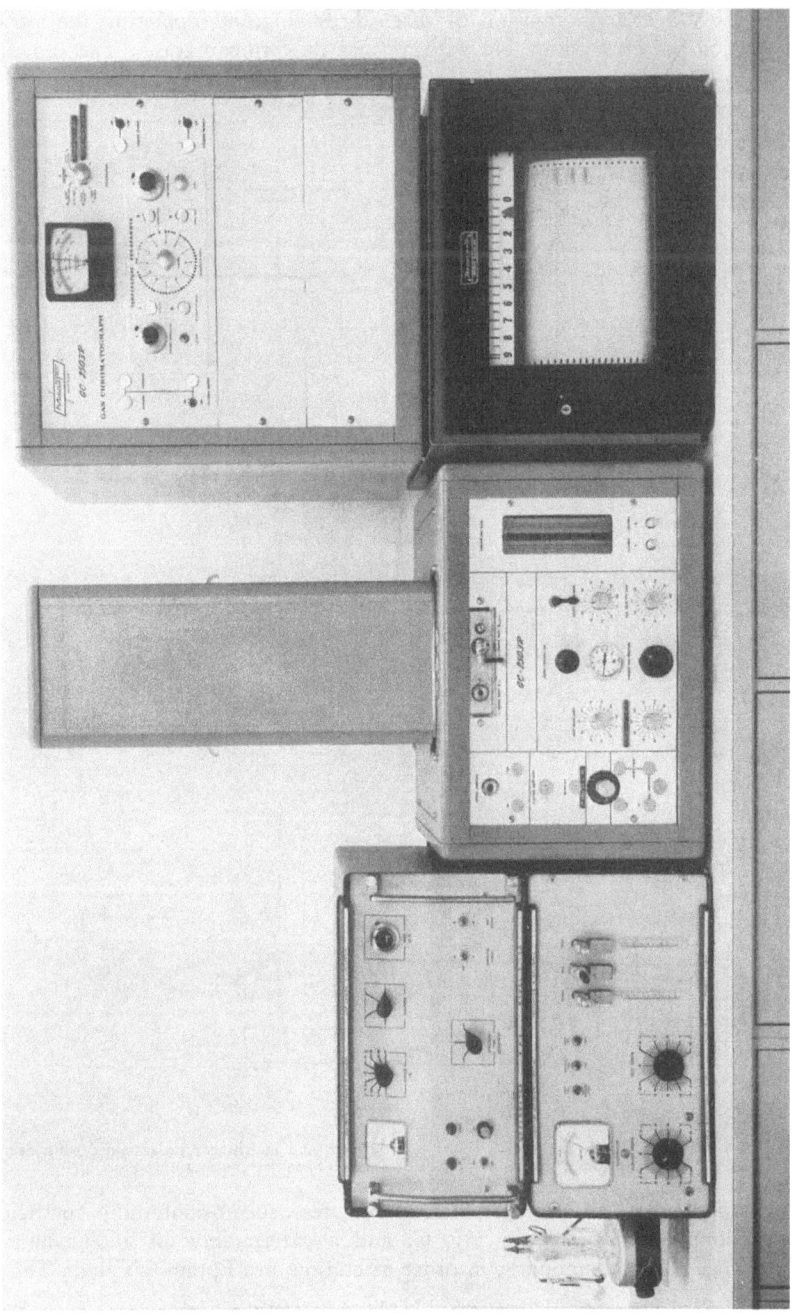

Fig. 5. Programmed Temperature Microcoulometric Gas Chromatograph

maximum sensitivity. The advantages of programming for chloride-containing pesticide compounds were presented by Bosin in 1962.

Fig. 6 is a photograph of three chromatograms depicting the various elution patterns achievable with a versatile chromatograph. The top chro-

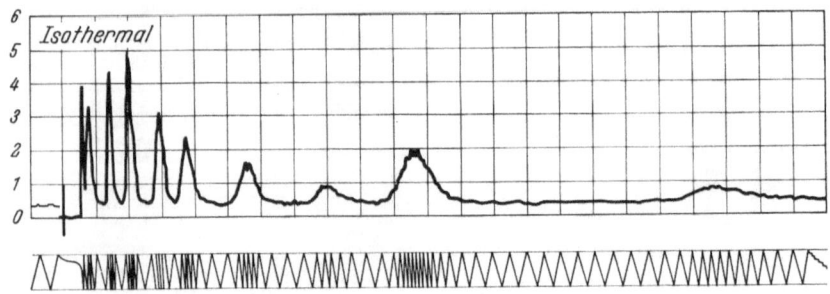

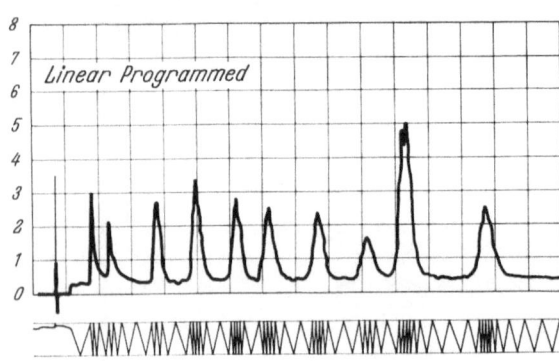

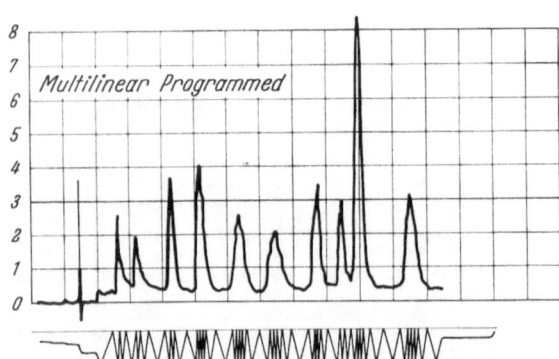

Fig. 6. Chromatograms: isothermal, linear programmed, and multilinear programmed runs. See text for identities of peaks of the same mixture of ten sulfur-containing pesticides

matogram is an isothermal run of ten sulfur-containing pesticides. Column conditions were 185° C. and a carrier flow of 100 cc./min. of nitrogen. The compounds, in order of elution, are Eptam [1], Tillam, Thimet,

---

[1] Pesticide chemicals mentioned in this text and their common or trade names are given in Table VI.

Dithiosystox, Ronnel, parathion, Thiodan I, Trithion, Thiodan II, and tedion. Total analysis time was 45 minutes. The middle chromatogram is a linear programmed run of the same mixture. Flow was 100 cc./min., initial temperature was 155° C., program rate was 3° C./min. Total analysis time was 27 minutes. The lower chromatogram is a multilinear programmed run of the same mixture. Initial temperature and flow were the same as the linear programmed run — namely, 155° C. and 100 cc./min. This program profile was isothermal for two minutes, program for two minutes at 15° C./min., isothermal for five minutes, program at 15° C./min. for six minutes, and isothermal for the remaining seven minutes. Total time, 22 minutes. Note the increased detectability on the last four compounds with only a slight loss in resolution between peaks 8 and 9. Also note the slight increase in resolution on the early eluters.

Because it is an absolute measuring instrument, the Microcoulometric Titrating System enabled us to begin an interesting study on column packings. Our object is to find a way to build a close approximation to the chimerically perfect column for gas chromatography of insecticides.

COULSON and DE VRIES (1959) had reported no significant differences in column performance when washing Chromosorb P with hydrochloric, sulfuric, or phorsphoric acids. Our preliminary work on halogenated pesticides indicated a significantly better yield on heptachlor, DDD, and DDT on Chromosorb P that had been sulfuric acid-washed. Our yields were 30 to 40 percent better on initial injections without "saturating" the column with those materials.

The procedure we used to evaluate washing of packing materials was to divide a three-pound lot of Chromosorb P into thirds and to place each portion in a borosilicate glass pan on a steam bath. Each portion was covered with a 6N-acid and heated for three hours with occasional stirring. The acids used were hydrochloric, sulfuric, and nitric. The pans were then removed from the heat and allowed to stand overnight. The acid was decanted and the Chromosorb P was rinsed with deionized water three or four times. The Chromosorb P was then placed in a 35-mm.×2.5-foot column. Deionized water was passed up through the column until the conductivity of the water flowing out the top was lower than that of ordinary distilled water and closely approximated that of the water inflow. The Chromosorb was then dried in a convection oven at 110° C. and sieved, retaining the 30/60 mesh material. Column packings containing 2.5, 5, and 10 percent-by-weight 12,500-centistokes Dow Corning 200 Oil were prepared by a modification of the technique described by CASSIL (1962).

Fifty grams of the Chromosorb was placed in a 15 mm. I. D. tube, and a solution of the silicone oil in ethyl acetate was passed through the Chromosorb. The system was closed to prevent evaporation and concentration. The column material was thoroughly wetted, allowed to drain, extruded into a pan, and the bulk of the ethyl acetate evaporated in the atmosphere at 40° C. The material was then dried at 110° C. in a convection oven. Solutions for coating the Chromosorb were prepared by diluting 7.5 g. of the silicone oil with 352.5 ml. of ethyl acetate for the 2.5 percent-by-weight coating material. The solution for the five percent material was 15 g. of oil and 345 ml. of ethyl acetate. Thirty grams of oil and 330 ml. of ethyl acetate prepared the solution for the 10-percent coated packing.

Table I. *Percentages of sulfur and chlorine in test compounds*

| Compound | Percent by weight | |
|---|---|---|
| | Sulfur | Chlorine |
| Diazinon . . . | 10.5 | — |
| Disyston . . . | 35.1 | — |
| Eptam . . . . | 16.9 | — |
| Ethion . . . . | 33.4 | — |
| Guthion . . . . | 20.2 | — |
| Malathion . . . | 19.4 | — |
| Methyl Trithion | 30.6 | 11.3 |
| Parathion . . . | 11.0 | — |
| Ronnel . . . . | 10.0 | 33.1 |
| Tedion . . . . | 9.0 | 39.8 |
| Thimet . . . . | 36.9 | — |
| Thiodan . . . . | 7.9 | 52.3 |
| Tillam . . . . | 15.8 | — |
| Trithion . . . . | 28.0 | 10.3 |
| VC 1-13 . . . . | 10.2 | 22.5 |

Homogeniety of material prepared in this manner was checked by taking two additional 95-g. batches of the dry packing material and coating them to make five percent-by-weight oil. The final dry weight was 99.5 and 100.0 g., respectively. Four cuts were taken from the material in the column, checking the bulk density and the performance of a mixture of six insecticides on columns made from each of the four cuts. No difference in retention times of any of the compounds was noted.

Four-foot columns of each of the nine different chromatographic packings were made. All columns were heat-conditioned at one time with a nitrogen flow of 120 cc./min. for each column at a temperature of 250° C. for 18 hours. The columns were then tested without further treatment with single injections of fifteen different sulfur-containing insecticides. The test solutions were 100 p.p.m. Injections were made with a 10-$\mu l$. Hamilton syringe. The amounts injected varied from 0.1 $\mu g$. to 0.9 $\mu g$. of pesticide, depending upon sulfur content and elution time.

Table I lists the percentages of sulfur and chlorine in the test compounds. Table II is a chart of the relative elution times of the test insecticides on

Table II. *Relative retention times of some sulfur-containing pesticides: Ronnel chosen as unity. Four-foot column, five percent-by-weight Dow 200 Oil (12,500 centistokes) on acid-washed Chromosorb P (30/60 mesh.); column temperature, 200° C.; carrier gas flow rate, 100 cc./min. of nitrogen; all transfer lines and injection block temperature, 225° C.*

| Pesticide | Ratio |
|---|---|
| Eptam . . . . . . . . | 0.23 |
| Tillam . . . . . . . . | 0.30 |
| Thimet . . . . . . . | 0.54 |
| Diazinon . . . . . . . | 0.67 |
| Di-Syston . . . . . . | 0.71 |
| Systox, thiono-isomer . | 0.71 |
| Systox, thiol-isomer . . | 0.71 |
| VC 1-13 . . . . . . . | 0.85 |
| Ronnel . . . . . . . . | 1.00 |
| Parathion . . . . . . | 1.08 |
| Malathion . . . . . . | 1.12 |
| Thiodan I . . . . . . | 1.59 |
| Methyl Trithion . . . . | 2.54 |
| Thiodan II . . . . . . | 2.58 |
| Ethion . . . . . . . . | 2.95 |
| Trithion . . . . . . . | 3.36 |
| Tedion . . . . . . . . | 5.80 |

a five percent-by-weight 12,500 centistokes Dow Corning 200 Oil column at 200° C. and 100 cc./min. of nitrogen. Relative elution times are based on Ronnel as unity. There are several reasons for this choice. First, it

Table III. *Determination of Tedion by halogen and sulfur methods: Tedion solution contained 10.0 mg. per 100 ml. (0.1 µg. per µl.)*

| Titration cell | Tedion solution injected, µl. | µg. | | Percent theoretical |
|---|---|---|---|---|
| | | Cl added | Cl found | |
| T-200-S . . . . . . | 1.7 | 0.067 | 0.066 | 98.5 |
| Halogen sensitive. . | 3.1 | 0.123 | 0.116 | 94.3 |
| Bias 250 mv. . . . . | 6.2 | 0.247 | 0.23 | 93.3 |

| Titration cell | Tedion solution injected, µl. | µg. | | Percent theoretical |
|---|---|---|---|---|
| | | S added | S found | |
| T-200-P . . . . . . | 0.5 | 0.0045 | 0.0042 | 93.4 |
| Sulfur sensitive. . . | 0.7 | 0.0063 | 0.007 | 111 |
| Bias 125 mv. . . . . | 5.0 | 0.045 | 0.046 | 102 |

Table IV. *Recovery data for evaluation of column packing material washed with various mineral acids: all columns, four-foot aluminum; inlet and transfer lines, 225° C.; carrier gas flow, 100 cc./min. of nitrogen. Recoveries are expressed as percent of theoretical*

| 6 N-acid | HCl | | | H₂SO₄ | | | HNO₃ | | |
|---|---|---|---|---|---|---|---|---|---|
| Percent-by-weight 12 500 centistokes Dow 200 Oil | 2.5 | 5.0 | 10 | 2.5 | 5.0 | 10 | 2.5 | 5.0 | 10 |
| Colum oven temperature | 175° C. | 190° C. | 210° C. | 175° C. | 190° C. | 210° C. | 175° C. | 190° C. | 210° C. |
| Column number | 1 | 4 | 7 | 2 | 5 | 8 | 3 | 6 | 9 |
| Test compound | | | | | | | | | |
| Eptam . . . . | 79 | 58 | 74 | 81 | 63 | 84 | 69 | 71 | 87 |
| Tillam . . . . | 55 | 52 | 68 | 67 | 54 | 80 | 60 | 80 | 85 |
| Thimet . . . . | 62 | 72 | 64 | 73 | 69 | 79 | 69 | 61 | 70 |
| Diazinon . . . | — | 85 | 76 | 94 | 78 | 82 | 78 | 78 | 71 |
| Di-Syston . . . | 79 | 84 | 83 | 76 | 80 | 82 | 78 | 67 | 84 |
| VC 1-13 . . . | 65 | 84 | 85 | 83 | 87 | 83 | 74 | 87 | 88 |
| Ronnel . . . . | 73 | 83 | 78 | 74 | 70 | 72 | 82 | 75 | 81 |
| Parathion . . . | 62 | 81 | 76 | 66 | 94 | 78 | 79 | 80 | 79 |
| Malathion. . . | — | 42 | 23 | — | 40 | 21 | 66 | 28 | 18 |
| Thiodan I. . . | 83 | 82 | 86 | 82 | 79 | 87 | 83 | 85 | 89 |
| Thiodan II . . | 80 | 82 | 72 | 82 | 82 | 83 | 76 | 79 | 84 |
| Methyl Trithion[a] | 30 | 56 | 49 | 46 | 52 | 48 | 68 | 44 | 57 |
| Ethion[a] . . . | 31 | 57 | 52 | 37 | 77 | 49 | 74 | 56 | 55 |
| Trithion[a] . . . | 59 | 72 | 72 | 66 | 73 | 69 | 80 | 65 | 75 |
| Tedion[a] . . . | 86 | 77 | 92 | 91 | 86 | 93 | 91 | 78 | 94 |
| Average column efficiency . . . | 65 | 71 | 70 | 73 | 72 | 73 | 75 | 69 | 74 |

[a] Column temperatures elevated 20° C. from indicated initial temperature to decrease analysis time.

exhibits good chromatographic behavior. Second, its actual elution time is median and close to that of aldrin, which has been chosen as unity by COULSON *et al.* (1960) and BURKE and JOHNSON (1962) when describing relative retention times. The third factor is that it contains both halogen and sulfur. Since the halogen determination is by precipitation, the efficiency of the column for a compound is easily determined. Thus, it is relatively easy to determine the optimum triiodide concentration for quantitative oxidation of the sulfur dioxide (formed when a compound passes through the micro-combustion furnace) to sulfur trioxide. The validity of the above reasoning is demonstrated in Table III which shows data for Tedion, which contains 39.8 percent-by-weight chlorine and nine percent-by-weight sulfur.

Table IV lists the columns tested, compounds used, and recovery data. Recoveries ranged from 52 to 94 percent, disregarding the data for mala-thion and Ethion, which showed extremely poor chromatographic behavior. The data are insufficient for a thorough statistical analysis. As a group, the nitric acid-washed material scored the highest with sulfuric acid-washed an extremely close second. Preliminary work indicates that low percentage coating on glass spheres in an all-glass system gives higher yields for malathion, confirming the findings of BECKMAN and BEVENUE (1963).

Table V. *Recovery data on the three five-percent-by-weight 12,500 centistokes Dow 200 Oil columns, following "chemical conditioning": column oven tempera-ture, 190° C.; flow, 100 cc./min. of nitrogen. Recoveries are expressed as percent of theoretical*

| Test compound | Column 4 (HCl-washed) | | Column 5 (H$_2$SO$_4$-washed) | | Column 6 (HNO$_3$-washed) | |
|---|---|---|---|---|---|---|
| | Before conditioning | After conditioning | Before conditioning | After conditioning | Before conditioning | After conditioning |
| Eptam . . . . | 58 | 78 | 63 | 75 | 71 | 82 |
| Thimet . . . . | 72 | 74 | 69 | 79 | 61 | 78 |
| Diazinon . . . | 85 | 82 | 78 | 87 | 78 | 91 |
| Di-Syston . . . | 84 | 81 | 80 | 86 | 67 | 84 |
| VC 1-13. . . . | 84 | 80 | 87 | 89 | 87 | 91 |
| Ronnel . . . . | 83 | 79 | 70 | 79 | 75 | 79 |
| Parathion . . . | 81 | 80 | 94 | 90 | 80 | 85 |
| Thiodan I[a] . . | 82 | 88 | 79 | 92 | 85 | 89 |
| Methyl Trithion[a] | 56 | 59 | 52 | 61 | 44 | 57 |
| Ethion[a]. . . . | 57 | 77 | 77 | 78 | 56 | 82 |
| Trithion[a] . . . | 72 | 81 | 73 | 82 | 65 | 80 |
| Tedion[a]. . . . | 77 | 91 | 86 | 93 | 78 | 93 |
| Average column efficiency | 74 | 79 | 76 | 83 | 71 | 83 |

[a] Column temperatures elevated to 210° C. to decrease analysis time.

Table V shows recovery data for the three five-percent oil columns following chemical conditioning with four injections of mixed pesticides. Each injection contained 100 $\mu$g. of Eptam, Thimet, diazinon, Di-Syston, VC 1-13, Ronnel, parathion, Thiodan I, Methyl Trithion, Ethion, Trithion, and Tedion. In an effort to minimize the effect of syringe error, the test solutions were diluted so that a 40-$\mu$l. injection was made for each test. Solutions of Eptam, Thimet, and Di-Syston contained 0.2 $\mu$g./40 $\mu$l.. All other solutions contained 0.8 $\mu$g./40 $\mu$l.

Future work will expand this study to include other supports, coatings, mixtures, tandem packings, washing techniques, and column construction materials in an effort to give the residue chemist that prized do-all column. We look forward to being able to describe a column which is extremely efficient for both halogen- and sulfur-containing pesticides. We eagerly solicit any comments or suggestions you may care to make to improve the state of the art.

System linearity has been checked using aldrin and Tedion for the chloride determination and Tedion for the sulfur determination. The aldrin was run on a well-conditioned four-foot, five-percent column, and the Tedion an a 10-inch, 10-percent column to shorten the analysis time.

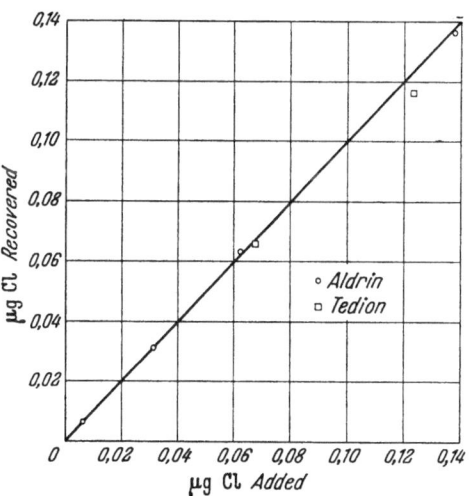

Fig. 7. System linearity: halogen determination

Fig. 7 depicts system linearity on the halogen determination. The drawn line is theoretical. The amounts depicted show excellent linearity from thousandths of a microgram to tenths of a microgram. The abscissa represents the amount added and the ordinate represents the amount recovered. Deviation of points are within sampling errors of microsyringes. Fig. 8 represents the same type data for the sulfur determination.

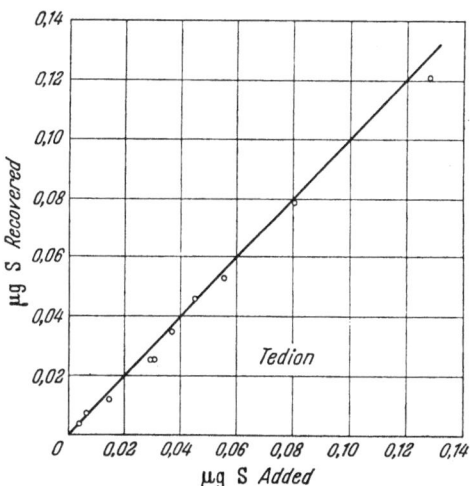

Fig. 8. System linearity: sulfur determination

## IV. Conclusions

In summary, we have described a system that will make almost any chromatograph a pesticide residue analyzer. In addition, it is a useful tool for optimizing conditions in residue analysis.

Table VI. *Common and chemical names of pesticides mentioned in text*

Common or trade-name designations of compounds mentioned in the text are listed alphabetically below, with the chemical name conforming to *Chemical Abstract's* nomenclature of the major ingredient.

| | |
|---|---|
| Aldrin. . . . . . | 1,2,3,4,10,10-hexachloro-1,4,4a,5,8,8a-hexahydro-1,4-*endo,* *exo*-5,8-dimethanonaphthalene |
| DDD . . . . . . | 2,2 bis-(*p*-chlorophenyl)-1,1-dichloroethane |
| DDT . . . . . . | 1,1,1-trichloro-2,2-bis(*p*-chlorophenyl) ethane |
| Diazinon®. . . . | *O,O*-diethyl-*O*-(2-isopropyl-4-methyl-6-pyrimidinyl) phosphorothioate |
| Di-Syston ® . . . | *O,O*-diethyl *S*-(ethylthio) ethyl phosphorodithioate |
| Eptam. . . . . . | Ethyl *N,N*-di-*n*-propyl thiolcarbamate |
| Ethion . . . . . | *O,O,O',O'*-tetraethyl *S,S'*-methylene bisphosphorodithioate |
| Guthion® . . . . | *O,O*-dimethyl *S*-(4-oxobenzotriazino-3-methyl) phosphorodithioate |
| Heptachlor. . . . | 1,4,5,6,7,8,8-heptachloro-3a,4,5,5a-tetrahydro-4,7-*endo*-methanoindene |
| Malathion . . . . | *O,O*-dimethyl *S*-1,2-di (ethoxy carbamyl) ethyl phosphorodithioate |
| Methyl Trithion® . | *O,O*-dimethyl *S-p*-chlorophenylethiomethyl phosphorodithioate |
| Parathion . . . . | *O,O*-diethyl *O-p*-nitrophenyl phosphorothioate |
| Ronnel ® . . . . | *O,O*-dimethyl *O*-2,4,5-trichlorophenyl phosphorothioate |
| Tedion ® . . . . | *p*-chlorophenyl-2,4,5-trichlorophenyl sulfone |
| Thimet ® . . . . | *O,O*-diethyl *S*-(ethylthiomethyl) phosphorodithioate |
| Thiodan ® . . . . | 6,7,8,9,10,10-hexachloro-1,5,5a,6,9,9a-hexahydro-6,9-methano-2,3,4-benzodioxathiepin-3-oxide (Isomers I and II) |
| Tillam. . . . . . | *n*-propyl *N,N*-ethyl-*n*-butyl thiolcarbamate |
| Trithion® . . . . | *O,O*-diethyl *S-p*-chlorophenylthiomethyl phosphorodithioate |
| VC 1-13 . . . . . | *O,O*-diethyl *O*-2,4-dichlorophenyl phosphorothioate |

## Summary

Investigations were made using the new Dohrmann Microcoulometric Titrating System as a total halogen or sulfur analyzer. Additional work involved its use as a separatory and analytical system when used in conjunction with the Micro-Tek Temperature Programmed Gas Chromatograph.

The course of this work on pesticide samples revealed performance differences in the gas chromatograph columns which were dependent upon the treatment of the solid supports. The absolute nature of the Microcoulometer allowed quantitative evaluation of column and system performance. The relative elution times of some thiophosphates have been charted using these columns.

System linearity checks were run down to the thousandths of a microgram level. Temperature programming proved valuable for samples containing several pesticides whose elution times differed considerably. Some very short columns were used as a rapid-screening method.

## Résumé *

Des études ont été faites avec le nouveau système à titrage microcoulométrique Dohrmann comme analyseur des halogènes totaux ou du soufre. Un travail supplémentaire entraîna son emploi comme système de séparation

---

* Traduit par R. MESTRES.

et d'analyse en l'utilisant conjointement au chromatographe Micro-Tek à température programmée.

Le développement de ce travail sur des échantillons de pesticides a révélé des differences de performances entre les colonnes provenant du traitement des supports solides. La microcoulométrie, grâce à ses résultats en valeur absolue, a permis une évaluation quantitative des performances de la colonne et de l'appareillage. Les temps de rétention relatifs de quelques thiophosphates, avec ces colonnes, ont été notés.

Des essais de linéarité de l'appareil ont été effectués jusqu'au niveau des millièmes de microgramme. La température programmée s'est montrée précieuse pour les échantillons contenant plusieurs pesticides dont les temps de rétention diffèrent considérablement. Des colonnes très courtes ont été utilisées comme méthode rapide de recherche préliminaire.

## Zusammenfassung *

Das neue mikrocoulometrische Titrationssystem nach Dohrmann zur Analyse von Gesamt-Halogen oder -Schwefel wurde untersucht. Weitere Arbeiten beschäftigten sich mit seiner Verwendung als Trenn- und Analysengerät in Verbindung mit dem temperaturprogrammierten „Micro-Tek"-Gaschromatographen.

Im Verlaufe dieser Arbeiten an Pesticidproben ergaben sich Verfahrensunterschiede mit verschiedenen Trennsäulen, die von der Behandlung des Trägermaterials abhängig waren. Die absolute Arbeitsweise des Mikrocoulometers erlaubt eine quantitative Ermittlung der Funktion der Trennsäule und des gesamten Systems. Die bei der Verwendung dieser Säulen erhaltenen relativen Elutionszeiten für einige Thiophosphate werden dargestellt.

Die Linearität des Systems wurde bis hinunter zu einem Tausendstel Mikrogramm geprüft. Die Temperaturprogrammierung erwies sich als wertvoll für Proben, die mehrere Pesticide enthalten, wenn deren Elutionszeiten stark differieren. Einige extrem kurze Säulen wurden für Schnelltests eingesetzt.

## References

BECKMAN, H., and A. BEVENUE: The effect of the column tubing composition on the recovery of chlorinated hydrocarbons by gas chromatography. J. Chromatog. 10, 231 (1963).

BOSIN, W. A.: Temperature programmed microcoulometric gas chromatograph for pesticide residue analysis. Amer. Chem. Soc., 143rd National Meeting, Atlantic City, September, 1962.

BURKE, J., and L. JOHNSON: Investigations in the use of the micro-coulometric gas chromatograph for pesticide residue analysis. J. Assoc. Official Agr. Chemists 45, 348 (1962).

CASSIL, C. C.: Pesticide residue analysis by microcoulometric gas chromatography. Residue Reviews 1, 37 (1962).

COULSON, D. M., and L. A. CAVANAGH: Automatic chloride analyzer. Anal. Chem. 32, 1245 (1960).

* Übersetzt von H. FREHSE.

COULSAN, D. M., L. A. CAVANAGH, J. E. DE VRIES, and B. WALTHER: Microcoulometric gas chromatography of pesticides. J. Agr. Food Chem. 8, 399 (1960).
—, and J. E. DE VRIES: Pesticide residues on fresh vegetables. Stanford Research
Institute Report No. 9, Technical Report No. III (1959).
—, A. S. HUENE, and L. A. CAVANAGH: Effects of pesticides on animals and
human beings. Stanford Research Institute Report No. 3, Technical Report
No. I (October 3, 1960).
ERICKSON, L. C., and H. Z. HIELD: Determination of 2,4-dichlorophenoxyacetic
acid in citrus fruit. J. Agr. Food Chem. 10, 204 (1962).

# Comparison of flame ionization and electron capture detectors for the gas chromatographic evaluation of herbicide residues

By

Hirsh S. Segal * and M. L. Sutherland *

With 9 figures

## Contents

## I. Introduction

This is a brief report on a comparison of the performance of flame ionization and electron capture detectors in the gas chromatographic analysis of a chlorinated herbicide in raw sugar cane juice. This work was originally undertaken to determine the level of any residue in sugar cane treated with Fenac, an isomeric mixture of polychlorinated phenylacetic acid. The active ingredient, which comprises almost half of the mixture, is 2,3,6-trichlorophenylacetic acid. It was principally to

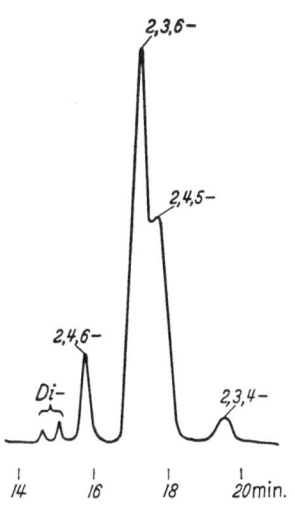

Fig. 1. Principal components of 10 μg. of Fenac, an isomeric mixture of polychlorinated phenylacetic acid. Column: 10 percent Ucon Polar on 70/80 Anachrom ABS, ¹/₄ inch × 3 feet; detector: flame; chromatograph: F & M 1609; program: 120° C. for 10 min., 42° C./min. to 195° C., hold; attenuation: 64 × 10

* Amchem Products, Inc., Ambler, Pennsylvania.

the determination of this isomer that we directed our efforts, while still keeping an eye on the rest of the mixture. Fig. 1 shows a GC trace of Fenac with isomers labeled.

## II. Sample preparation

Carbon-14 labeled Fenac was used to develop a satisfactory method of extraction and sample cleanup. The procedure arrived at was extraction of the acidified cane juice into carbon tetrachloride. From there it was extracted into dilute sodium bicarbonate which was acidified for re-extraction into carbon tetrachloride. The final extract was evaporated to dryness and esterified with an ether solution of diazomethane. The esterified sample was taken to dryness and picked up in an appropriate solvent depending on the detector employed.

## III. Flame ionization

The early work was done on an F & M Model 500 gas chromatograph fitted with their flame ionization detector attachment. The column was $1/4$ inch$\times$3 feet, 10 percent Ucon Polar on 70/80 mesh Anachron ABS. The

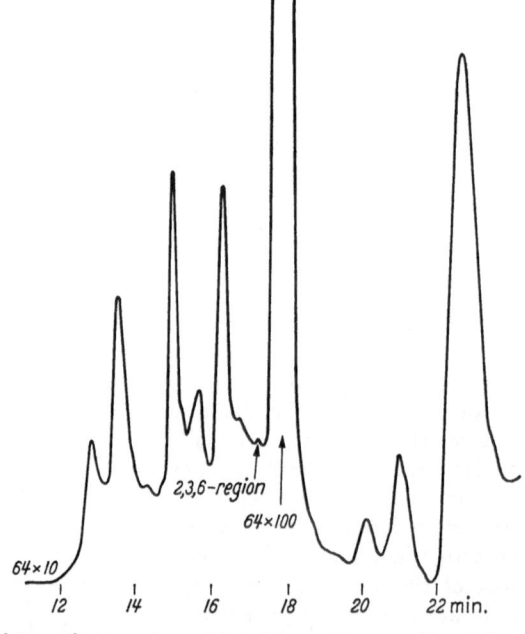

Fig. 2. Extract of 50 g. of untreated cane juice. Column: 10 percent Ucon Polar on 70/80 Anachrom ABS, $1/4$ inch $\times$ 3 feet; detector: flame; chromatograph: F & M 1609; program: 120° C. for 10 min., 42° C./min. to 95° C., hold; attenuation: 64 $\times$ 10

effective limit of detection for a 50 g. sample was 0.5 $\mu$g. or 10 parts per billion (p.p.b.) of Fenac. Fig. 2 shows the final extract from 50 g. of cane juice on 10 percent Ucon Polar Column, flame ionization. A small interfering peak occurred at the same retention time as the 2,3,6-isomer and was equal to perhaps five p.p.b. The retention time for the 2,3,6-isomer is surrounded by very large peaks and the detection of such traces in the presence of these nearby giants is not a happy situation. The slightest deterioration in column efficiency will obliterate such a precarious determination and of course any further increase in sensitivity is precluded.

A two percent Ucon Polar column improved resolution somewhat but the nearby large peaks were still uncomfortably close.

A ¹/₂ percent Ucon Polar column on the same support allowed a much lower operating temperature. A mild temperature rise was fitted into the program and the resolution improved considerably. While the neighboring 2,3,6- and 2,4,5-trichloro-isomers had been barely separated on the 10 percent column, i. e., 0.3 minutes between peaks, and only one minute on the 2 percent column, here they are five minutes apart. Fig. 3 shows A, 0.14 μg.

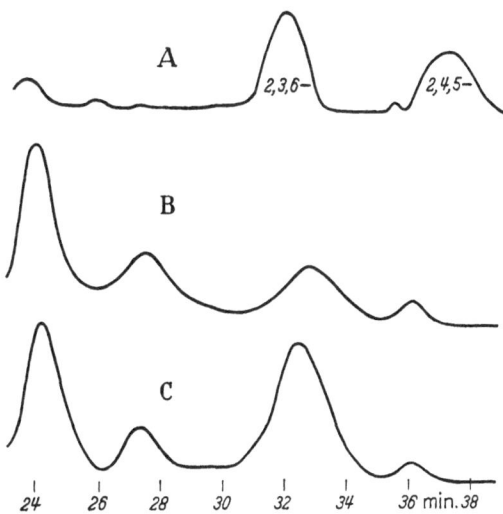

Fig. 3. Limits of flame ionization sensitivity: A, 0,14 μg. of Fenac; B, 50 g. of check cane; C, 5 p.p.b. recovery. Column: ¹/₂ percent Ucon Polar on 70/80 Anachron ABS, ¹/₄ inch × 3 feet; detector: flame; chromatograph: F & M 1609; program: 75° C. for 6 min., 3° C./min. to 95° C., hold; attenuation: 2 × 1

of Fenac on the ¹/₂ percent Ucon Polar column. A 50-g. check cane sample shows an interference of about two p.p.b. and no large peaks in the near vicinity of the 2,3,6-isomer (Fig. 3, B). The nearest large peak is more than eight minutes prior to the area of interest. The large neighboring peaks seen previously are nowhere in this picture and emerge only when the column is cooked out at 150° C. This relative freedom from interferences and low bleed afforded by this column have allowed us to fully utilize the sensitivity of this detector. While on the 10 percent column we were held to an attenuation of 640 x, on the ¹/₂ percent column we have reduced the attenuation to 2X. The five p.p.b. recovery requires a blank crop correction and yields a 60 percent recovery (Fig. 3, C). The extreme limit of detection here, with corrections for crop blanks is two p.p.b. In practice, a number of samples had no interference in this region and the practical detection limit improved to less than one p.p.b.

## IV. Electron capture

It was at this point that Dr. L. R. Mattick, New York State Agricultural Experiment Station, Cornell University, demonstrated to us the applicability of the electron capture detector to this problem. Without resorting to temperature programming or very long retention times at low temperatures, Mattick easily detected a nanogram of Fenac in an extract equal to a gram of cane juice and thus achieved a sensitivity of one p.p.b. (see Fig. 4). A number of crop samples were run and showed no interferences in the region of interest. Nevertheless, it was clear to see that there were other substances present, sensitive to the electron capture detector, which could have provided serious interference to a compound of somewhat different retention time.

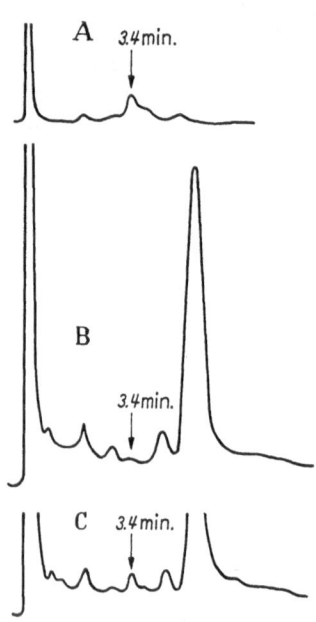

Fig. 4. Electron capture at 1 p.p.b.: *A*, 1.3 ng. of Fenac; *B*, 1 g. of check cane juice; *C*, 1 p.p.b. fortification. Column: 20 percent silicone grease on 80/100. Chromosorb-W, ¼ inch × 2 feet; detector: electron capture; chromatograph: Research specialities; temperature: 153° C.; carrier: nitrogen, 172 cc. per min.

## V. Flame ionization versus electron capture

### a) Requirements

With different column packings, temperatures, carrier gas flow rates, etc., we were unable to make any but the crudest comparison of results with the two detectors. It was decided to make a careful comparison, in our laboratory, of the flame ionization and electron capture detectors using a single column and set of conditions held as rigidly fixed as possible in order to eliminate all variables except those inherent in the detectors themselves.

### b) Instrumentation and conditions for comparison

A Wilkens Hy-Fi chromatograph was used since both electron capture and flame ionization detectors were interchangeable in this unit without changes in operation conditions. Only one compromise had to be made, and that was for carrier gas flow. Optimum sensitivity for this electron capture detector is obtained at a high gas flow, in the neighborhood of 150 ml./min. By contrast, the flame detector will not support a flame with such a carrier gas flow and indeed operates best at 30 to 40 ml./min. A compromise was struck at 92 ml./min. which was the highest flow at which even an oversized,

somewhat noisy flame would not blow out. The column was a one percent
Ucon Polar on 70/80 mesh Anachrom ABS, ¹/₈ inch×5 feet. Since there is
roughly a 1,000-fold difference in sensitivity for this compound when
comparing these detectors, aliquots of the same samples in a 1,000 : 1 ratio
were chosen for work with the flame and electron capture detectors, respecti-

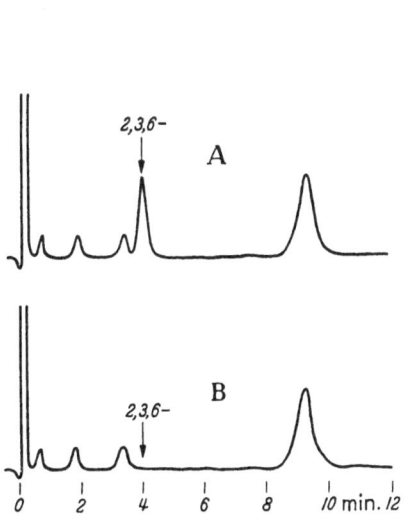

Fig. 5. Fortified (A) and check (B) cane by electron capture; 0.2 p.p.m. in 10 mg. Column: 1 percent Ucon Polar on 70/80 Anachron ABS, ¹/₈ inch
× 5 feet; detector: electron capture; chromatograph: Wilkens Hy-Fi; temperature: 137° C.;
carrier nitrogen, 92 cc./min.

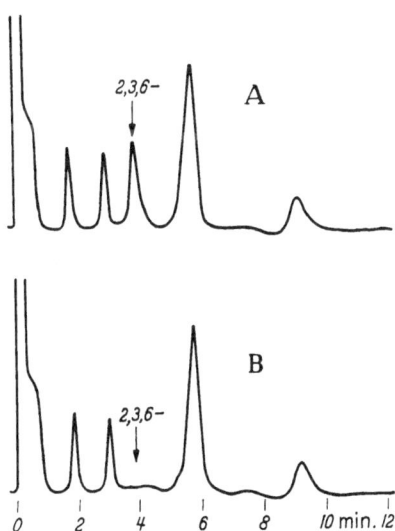

Fig. 6. Fortified (A) and check (B) cane by flame
ionization: 0.2 p.p.m. in 10 g. Column: 1 percent
Ucon Polar on 70/80 Anachrom ABS, ¹/₈ inch
× 5 feet; detector: flame ionization; chromatograph: Wilkens Hy-Fi; temperature: 137° C.;
carrier: nitrogen, 92 cc./min.

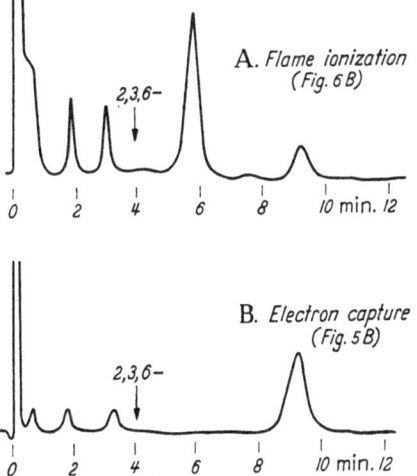

Fig. 7. Check cane: flame ionization versus electron
capture

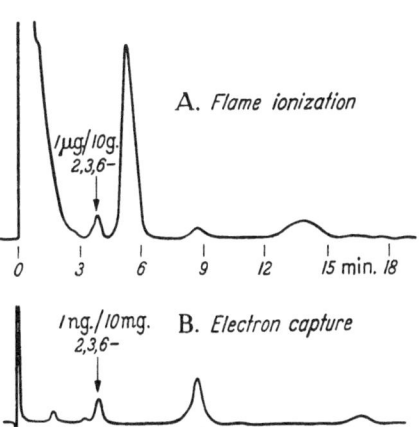

Fig. 8. Fortified cane: flame ionization (A) versus
electron capture (B). Column: 1 percent Ucon
Polar; temperature: 137° C.

vely. For the electron capture detector a 10-mg. untreated crop sample was followed by an identical aliquot fortified at 0.2 p.p.m. (see Fig. 5). After these runs, the detectors were exchanged, the air and hydrogen turned on for the flame ionization detector and 10-g. aliquots of the same samples were repeated (see Fig. 6). A similar comparison results, with some minor variations, from an untreated crop from another sugar cane growing area (see Fig. 7 and Fig. 8).

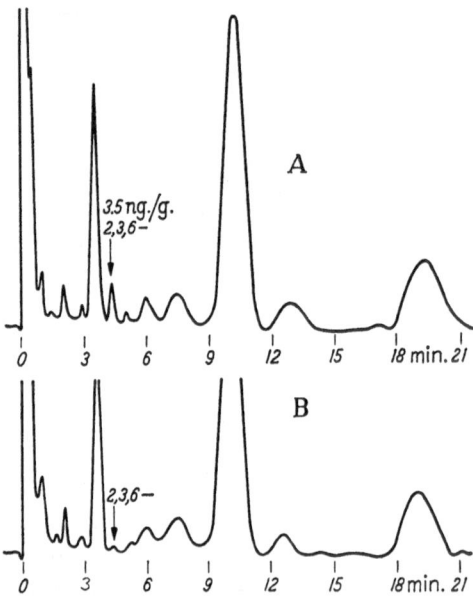

*c) Results and comparisons*

As expected, some of the peaks seen with the flame "disappear" when examined with the electron capture detector. However, one peak actually increased in size and several smaller peaks appear where none were seen before. One may assume that minute quantities of naturally occurring, very active, electron-capturing substances account for these new peaks.

Fig. 9. Electron capture in trouble: *A*, 3.5 p.p.b. fortification of 1 g. of cane; *B*, 1 g. of cane. Column: 1 percent Ucon Polar; temperature: 135° C.

In these curves, all peaks have been normalized to yield equal response for the 2,3,6-isomer at the 0.2 p.p.m. level, using appropriate standard amounts of this isomer as a guide. Under these same conditions, an attempt to approach a limit of detection of two p.p.b. with the electron capture detector and a one-gram sample brings us uncomfortably close to the tail from the preceding peak (see Fig. 9). Here we are in much the same type situation as we were in the earlier work with the flame, i. e., column manipulations must be employed to separate interferences from the compound of interest. It can be seen from this analysis that the election capture detector will favor certain problems while in other cases the flame ionization detector will be the better choice. It therefore behooves the analyst attacking a new problem to investigate both detectors.

## VI. Special advantages of electron capture

Aside from its relative specificity, there are two clear advantages to using the electron capture detector, both related to its high sensitivity. The sample size required is considerably smaller than with the flame and consequently the sample preparation may be simplified. Again, even if a modest size sample is prepared it may be divided into dozens of identical portions which can than be used in searching for the best gas chromatographic column and conditions.

It seems clear that the electron capture detector is a welcome addition to the tools of the research analyst, but — like so many before — it brings its own peculiarities and drawbacks with it.

## Summary

Employing appropriate sample preparation and gas chromatographic conditions both flame ionization and electron capture detectors are capable of low parts per billion residue detection for a polychlorinated herbicide in sugar cane juice.

A rigorous comparison of these two detectors reveals that each has its own special advantages when dealing with o group of plant extract interferences and that it behooves the residue analyst to investigate the use of both detectors when undertaking a new problem.

## Résumé *

Grâce à une préparation convenable de l'échantillon et à des conditions opératoires particulières, les détecteurs à ionisation de flamme comme à capture d'électrons sont capables de déceler des résidus d'un herbicide organochloré à des doses pouvant atteindre quelques parties par milliard dans des jus de canne à sucre.

Une comparaison rigoureuse de ces deux détecteurs montre que chacun d'eux a ses propres avantages particuliers vis à vis d'interférences d'extraits végétaux. Il convient à l'analyste d'essayer ces deux détecteurs lorsqu'il envisage un problème nouveau.

## Zusammenfassung **

Durch geeignete Vorbehandlung und unter entsprechenden gaschromatographischen Versuchsbedingungen können mit Flammenionisations- und Elektroneneinfang-Detektoren Rückstände von polychlorierten Herbiciden in Zuckerrohrsaft in der Größenordnung von wenigen ppb (parts per billion) bestimmt werden.

Beim gründlichen Vergleich beider Detektortypen wird deutlich, daß jeder seine speziellen Vorteile hat, wenn man eine bestimmte Gruppe pflanzeneigener Störstoffe untersucht. Der Rückstandsanalytiker muß die Brauchbarkeit beider Typen prüfen, wenn er sich vor ein neues Problem gestellt sieht.

---

* Traduit par R. Mestres.
** Übersetzt von H. Frehse.

# Applications of polarography for the detection and determination of pesticides and their residues

By

Raymond J. Gajan *

With 7 figures

## Contents

## I. Introduction

Professor Jaroslav Heyrovsky (1956), inventor of polarography and 1959 Nobel prize winner, defines "polarography" as the science of studying the processes occurring at the dropping mercury electrode. He further limits the term "polarography" to the capillary mercury electrodes, i. e., the dropping mercury electrode or streaming mercury electrode, because of their unique property of giving exactly reproducible results.

For the purpose of this paper oscillographic polarography is defined as any process whereby the polarographic phenomena occurring at the capillary mercury electrode are recorded on a cathode ray oscilloscope.

## II. Oscillographic polarography

The cathode ray oscilloscope was first used in polarography by Matheson and Nicols (1938). About the same time a cathode ray polarograph was also described by Müller et al. (1938).

Two radically different techniques are used in applying the cathode ray oscilloscope in polarography. The first technique is commonly referred to as the single sweep method. In single sweep instruments, such as that

---

* Division of Food, Food and Drug Administration, Department of Health, Education, and Welfare, Washington 25, D. C.

described by Reynolds and Davis (1953), the entire change of potential is effected during the life of a single drop. The resulting "polarotrace" is observed on a cathode ray oscilloscope. In order to obtain reproducible results the sweep is synchronized with the dropping rate and is confined to the final stages of the mercury drop when its growth rate is smallest. The drop time is usually seven seconds. The sweep time is set at two seconds, leaving a delay period of five seconds during which each new drop is forming. The rate of change of cell potential is fixed at 0.3 volt/second. The starting potential may be varied from +0.5 volt to −2.0 volts. By modifying the circuitry, it is also possible to use this type of instrument for derivative polarography in which the rate of change of the direct wave,

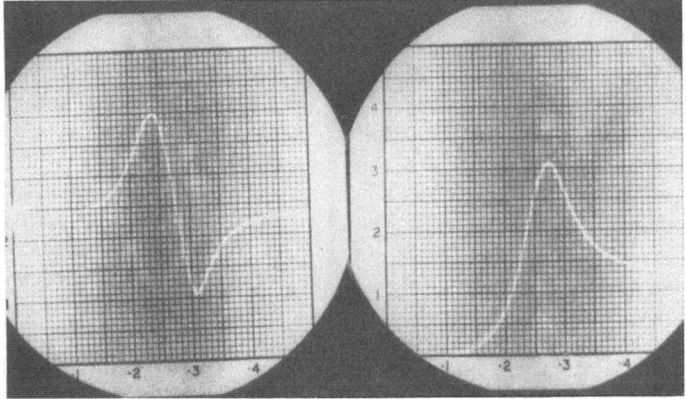

Fig. 1. Left: Derivative polarotrace of parathion, 20 µg. Right: Regular wave

$\frac{di}{dE}$ vs E, is measured instead of $i$ (current) vs E (potential). Fig. 1 shows examples of a derivative wave and a regular wave. The main advantages in using derivative polarography are that it will resolve two direct peaks which are close together and that it makes it possible to measure the concentration of a trace element when it is preceded by the reduction of a major constituent. Fig. 2 shows an example of the resolving power of the derivative circuit. In using this derivative circuit sensitivity is sacrificed by a tenfold factor.

The second oscillo-polarographic technique was first developed by Heyrovsky and Forejt (1943) and is referred to as the multisweep method, or oscillographic polarography with alternating current. They devised a polarograph which produced a potential-time relationship or its derivative, $\frac{dt}{dE}$ vs E. In this instrument an alternating current from a 50-cps source at constant amplitude is regulated to change the dropping mercury electrode from zero volt to −2.0 volts in a hundredth of a second and then back to zero volt in the next hundredth of a second. The frequency of the time sweep is synchronized with the applied alternating current voltage to produce a stationary potential-time figure on the oscilloscope.

Any process which results in a flow of current at the capillary mercury electrode produces a horizontal deflection on the potential-time curves. Thus, with this polarograph the cathode ray oscilloscope indicates depolarization by kinks or time lags on the curves. Each depolarizer, i.e., the component which reacts at the capillary mercury electrode, shows two kinks, one on the cathodic portion of the wave and the other on the anodic portion. The derivative wave of the potential-time curve, $\frac{dE}{dt}$ vs $E$, is usually recommended for analytical purposes. Heyrovsky and Forejt (1943) also used a streaming mercury electrode which corrects for the complications arising from the periodically changing drop area. Fig. 3 illustrates a typical oscillogram obtained with a multisweep instrument.

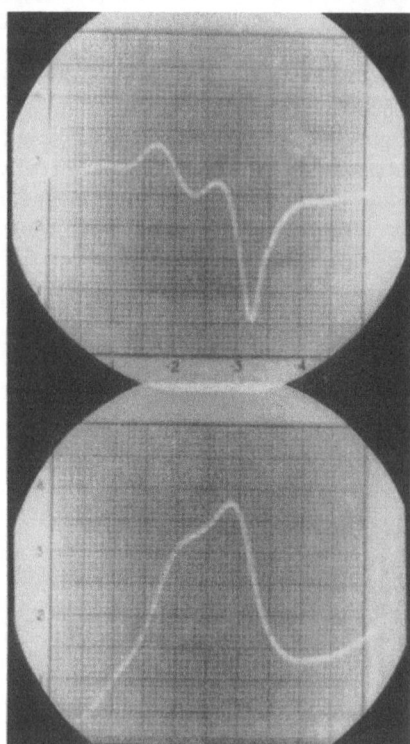

Fig. 2. Mixture of 20 µg. each of Thiono- (—0.60 volt) and Thiol-Systox (—0.68 volt). Top: Derivative trace. Bottom: Regular trace

The horizontal deflection of the cut of such a polarogram denotes the depolarization or half-wave potential, and the vertical deflection or depth of the cut denotes the diffusion current which increases proportionally with the concentration of the depolarizer. Because of widely different operating conditions multisweep instruments are able to detect many reactions at the capillary mercury electrodes which are not measurable by conventional polarography. For example, Heyrovsky (1957) was able to distinguish between o-, p-, and m-nitrophenols in the presence of nitrobenzene.

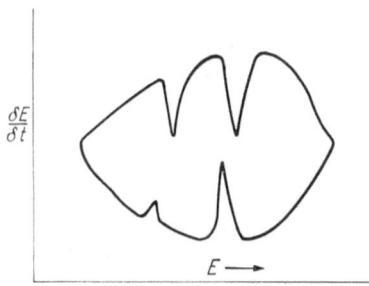

Fig. 3. Typical derivative oscillogramm, obtained with multisweep type instrument

Another interesting phenomenon specific for oscillographic polarography with the alternating current is the formation of artifacts. Heyrovsky (1956 b) reports that when vapors of carbon disulfide are polarographed, they give on the derivative potential-time curves, $\frac{dE}{dt}$ $vs\ E$, the cut-ins for sulfur anions or hydrogen sulfide. He thinks that the carbon disulfide is reduced at negative potentials above $-1.5$ volts to sulfur anions which are depolarized at $-0.6$ volt. In this case the artifact is the sulfur anion. When the direct current was adjusted so that it would not reach $-1.5$ volts, the sharp cut due to the artifact disappeared.

For qualitative analysis and kinetic studies, oscillographic polarographs with alternating current (the multisweep-type instrument) are preferred. The main reason for this is that their resolving power is greater and that many more substances can be polarographed.

However, for quantitative analysis the single sweep-type polarographs are recommended because of their greater sensitivity and accuracy. Also, they are not subject to errors caused by variations in drop size and/or too many sweeps applied to the same drop. These are matters of definite concern in using a multisweep-type instrument.

## III. General procedures followed in developing a polarographic method of analysis for pesticides

A majority of pesticides in use today are organic compounds.

Each group to be polarographed requires individual attention with regard to the selection of solvents, electrolytes, sample preparation, etc. In inorganic polarography most of these factors have been standardized to a great extent.

We usually proceed in the following manner when developing a polarographic method for the detection and determination of pesticide residues:

First, we examine the chemical structure of the pesticide for the presence of oxidizable or reducible groups such as the nitro group, halogens, carbonyl groups, etc. We also consider the possibility that, although the compound itself may not contain such a group, perhaps a derivative of it can be made that will. In this work breakdown and hydrolysis products have also been used.

Second, we search the literature for any information about the polarography of these groups. The most valuable single reference source for such information is the book entitled "Polarography in Medicine, Biochemistry, and Pharmacy", by BREZINA and ZUMAN (1958). The next source we try is the excellent series of reviews on organic polarography by WAWZONEK (1949, 1950, 1952, 1954, 1956, 1958, 1960, 1962). From these and other sources we usually find several candidate solvents and base electrolytes along with much pertinent data such as pH, temperatures, volumes, reference electrodes, etc., so essential to polarographic methodology.

The candidate solvents and electrolytes are then tested in the laboratory under various conditions and concentrations. Because of the rapidity of the cathode ray polarograph many combinations can be examined daily and a great deal of information rapidly collected. From these experiments we are usually able to select a solvent and base solution which most nearly meet previously determined specifications. In general, the solvent, to be usable in polarography, must be miscible with water, although completely nonaqueous solutions may be used. The solvent usually has an effect on the diffusion coefficient. It may also influence the surface tension and thus the drop time. Acetone, acetonitrile, dimethylformamide, ethanol, methanol, pyridine, dioxane, glacial acetic acid, ethylene glycol, and the ethylene ethers are usable solvents. To these solvents a suitable salt, or electrolyte, is added to carry the electrical current. The tetra-alkylammonium salts and the salts of the alkali metals — sodium, potassium, and lithium — are most commonly used.

Having selected the base solution and operating procedures we next try the proposed method on various plant extracts with and without known amounts of pesticide in order to determine what kind of sample preparation and cleanup, if any, are necessary for residue determinations.

With the solution of these problems we now have a method ready for verification and collaboration in the usual manner applicable to all instrumental methods.

The following precautions are essential in polarographic investigations of pesticide residues. All chemicals must be checked for purity and, if necessary, redistilled or recrystallized. This is especially true of solvents and column packings used in cleanup. Many of the interferences encountered in polarographic investigations have been traced to these two sources. Clean mercury and glassware are essential.

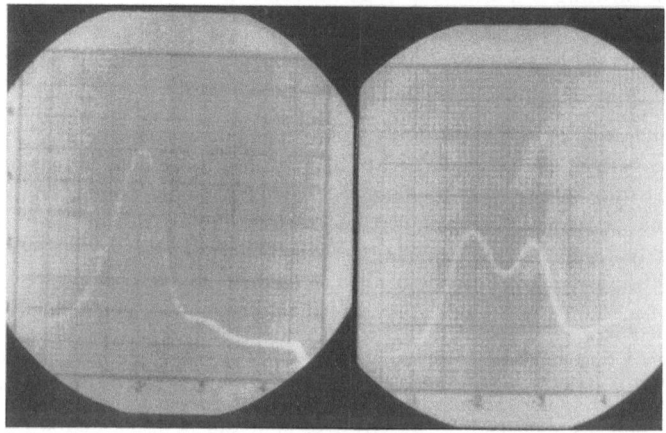

Fig. 4. Degradation of 50 μg. of Thiol-Systox in ethyl acetate. Left: Fresh solution. Right: Week-old solution

We have found that in very dilute solutions, 10 μg. to 50 μg. per ml., most of the pesticides studied are quite unstable. It is therefore advisable to use fresh reference solutions and to check them frequently for decomposition. Fig. 4 shows an example of the degradation of thiol-Systox in ethyl acetate at room temperature.

## IV. Applications of polarography to the determination of pesticide residues

Time does not permit a detailed comprehensive review of all the recent applications of oscillographic polarography to pesticide residue analysis. MARTENS and NANGNIOT (1963) have written an extensive review for *Residue Reviews*. However, I would like to call attention to several very recent studies which illustrate the versatility of this technique.

SOHR (1962) studied the polarography of several organophosphorus compounds, using an oscillographic polarograph with alternating current, a multisweep instrument. He investigated a series of trialkyl phosphates, triphenyl phosphate, a series of trialkyl phosphites, trimethyl thiophosphate, and benzyldimethylthiophospate. He used 1 $M$ potassium chloride in water and 1 $M$ potassium chloride in 50 percent methanol as base solutions, the latter with those compounds which were insoluble in the 1 $M$ potassium chloride solution. All of the compounds studied gave usable cathodic cuts on the derivative potential-time curve, $\frac{dE}{dt}$ $vs$ $E$.

The concentrations of the above compounds were in the $10^{-2}$ to $10^{-4}$ $M$-range. SOHR (1962) also found that the phosphoric acid esters and the thiophosphoric acid esters gave two cathodic cuts, the second cut due to an artifact caused by the formation of mercury complexes at the dropping mercury electrode.

BATES (1962) describes a method for the determination of "azinphos-methyl" (Guthion) residues in apples, pears, cucumbers, and tomatoes using a cathode ray polarograph, the Polarotrace, a single sweep-type instrument. He used a base solution containing 0.05 $M$ potassium chloride and 0.1 $N$ acetic acid in a 60 percent vol./vol. acetone-water mixture. Usable peaks were obtained at $-0.83$ volt $vs$ a standard calomel electrode, S.C.E. The method is reasonably specific since it is based on the reduction of the carbonyl group of the whole molecule. However, it will not distinguish between Guthion and its ethyl-analog. Interfering substances were removed on a column of magnesium oxide. The pesticide was eluted from the column with benzene. The average recovery of Guthion from the crops studied was 95 percent with a standard deviation of $\pm 12.2$ percent at the 0.5- and 1.0-p.p.m. range. The limit of detection was 0.1 $\mu$g. of Guthion per ml. of polarographic solution. This was equivalent to 0.1 p.p.m. for the crops and conditions used. However, since the polarographic measurements were made at sensitivity settings of 0.4 and 0.25 and the highest sensitivity of the instrument is 0.004, BATES (1962) suggests that much smaller amounts of Guthion could be detected with a more refined cleanup. He also suggests that a better cleanup may be necessary before the method can be applied to other crops, especially those containing high concentrations of plant pigments and waxes.

DAVIDEK and JANICEK (1961) used a new approach for the polarographic determination of DDT. They nitrated DDT with a nitration mixture of fuming nitric acid and concentrated sulfuric acid for ten minutes at 90 to 95° C. After cooling, this mixture was diluted with water and methyl alcohol so that the final mixture was 20 percent nitration mixture and 50 percent methanol. The half-wave potential of the tetranitro-derivative of DDT in this solution was $-0.13$ V $vs$ S.C.E.

According to DAVIDEK and JANICEK (1961) the advantages of this method over the colorimetric determination lie in the possibility of using an acid medium directly. They also claim it surpasses all other methods for DDT because of its high sensitivity and ease of operation. They used a conventional polarograph in their work. This method was included to illustrate an example of a new trend in polarography—that of polarograph-ing the nitration or nitrosation products of various compounds for the determination of the parent compound. Other types of derivatives may perhaps prove useful in this way. Fig. 5 illustrates waves obtained from nitro-derivatives of DDT and methoxychlor. Fig. 6 illustrates an example of a nitroso derivative of the organic compound, diethylstilbestrol.

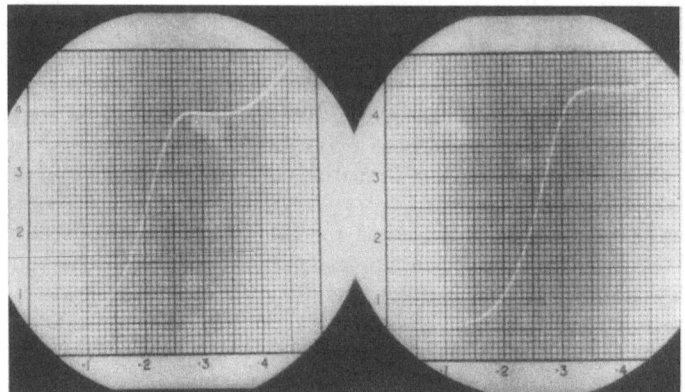

Fig. 5. Nitro-derivatives. Left: 20 μg. of p,p'-DDT. Right: 20 μg. of methyloxychlor

NANGNIOT (1960 a, b, and c) in Belgium has applied oscillographic polarography to the determination of sulfur, TMTD, ferbam, zineb, and ziram residues. He claims that they can routinely run a TMTD residue

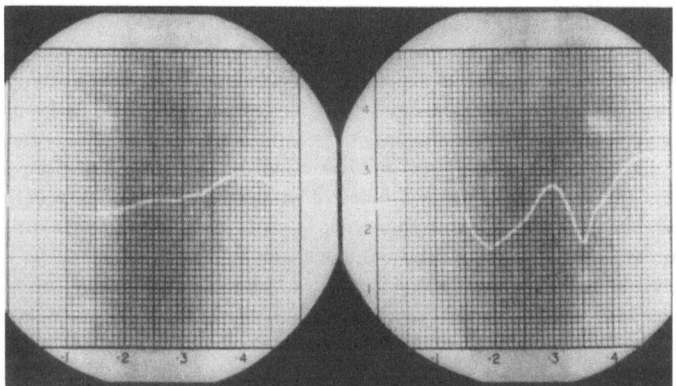

Fig. 6. Nitrosation. Left: Blank feed. Right: Feed and nitroso-derivative of diethyl-stilbestrol, 0.22 μg./ml.

determination or a sulfur residue determination in less than 12 minutes. They allow three minutes for extraction, two minutes for preparation of the polarographic solution, five minutes for deoxygenation, and two minutes for recording the polarogram. In his studies with TMTD he compared results by several polarographs. He was able to detect 35 μg. of TMTD/25 ml. by conventional polarography and 2 μg. of TMTD/25 ml. by using a single sweep oscillographic polarograph.

Publication is now pending (GAJAN 1963) on a rapid screening procedure we have developed for the detection and determination of parathion residues on some fruits and vegetables. It is capable of detecting less than 0.1 p.p.m. of parathion. A buffered solution containing 0.2 $M$ acetic acid

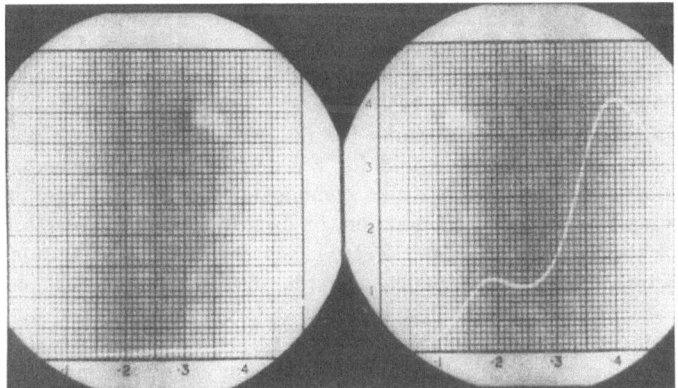

Fig. 7. Right: Mixture of 50 $\mu$g. each of parathion (—0.60 volt) and $p$-nitrophenol (—0.80 volt). Left: Blank solution

and 0.2 $M$ sodium chloride, and adjusted to pH of 5.0 with glacial acetic acid, was the supporting electrolyte used. The peak potential of parathion in reference to a silver wire electrode was $-0.62 \pm 0.2$ volt. $p$-Nitrophenol, one of the major breakdown products of parathion, has a peak potential of $-0.80$ volt in this base solution and thus does not interfere with the method. Fig. 7 shows the polarotrace of a mixture of parathion and $p$-nitrophenol. These peak potentials compare favorably with the half-wave potentials of these compounds determined in a similar base solution by MARTENS et al. (1961). Methyl anthranilate does not interfere. Twenty-five micrograms of parathion was determined in the presence of 1.2 mg. of methyl anthranilate.

Green beans, tomatoes, spinach, apples, brussels sprouts, and broccoli were the commodities studied. Briefly, the parathion was extracted from crop samples with acetonitrile following the procedure outlined by GETZ (1962). The acetonitrile was evaporated and the parathion was extracted from the aqueous residue with ethyl acetate. An aliquot of the ethyl acetate solution was taken just to dryness, the residue was dissolved in acetone, and an equal amount of electrolyte was added. This solution was then transferred to a polarographic cell and polarographed in the usual manner. The average recoveries from 68 samples of uncleaned extract, ranging from 0.1 to 5.0 p.p.m. averaged 91 percent with an average deviation of 7.6 percent. Since an occasional large interference wave has been observed, it is important to make another run on a cleaned up extract when any significant wave is detected. Whenever this occurred the sample extract was cleaned up on a charcoal column described by GETZ (1962), and the interference was usually eliminated. When such waves persist, verification of results by the colorimetric method of AVERELL and NORRIS (1948) or by the paper chromatography method of GETZ (1962) is suggested. Results obtained from field-sprayed commodities and those purchased in the supermarket compared favorably with results obtained by the Association of Official Agricultural Chemists colorimetric method of AVERELL and NORRIS (1948) and by paper chromatography using the GETZ (1962) procedure. In these studies a single sweep-type instrument, the Polarotrace, was used.

Various investigators, including KEMULA et al. (1952, 1956 a, 1956 b, 1956 c, 1960) and SANDI (1959), have successfully combined column chromatography and polarography for the detection and determination of the various isomers of DDT, and for the determination of parathion and its analogs. We have successfully combined paper chromatography and polarography for the determination of Systox, Thimet, Di-syston, and some of their analogs (GAJAN 1962). The pesticide sample solutions were first chromatographed on paper and the resulting spots were cut out, eluted with an electrolyte solution, and polarographed. Paper chromatography and thin layer chromatography in combination with polarography may well be a powerful tool in pesticide residue investigations.

Other pesticides and fungicides we have been able to detect and determine in residue quantities are PCNB, DDT, TCNB, Guthion, Trithion, malathion, and Dipterex. The last two pesticides were determined by using very slight modifications of the methods published by JURA (1955) and by GIANG and CASWELL (1957) several years ago.

## V. Conclusions

Oscillographic procedures are common practice in many fields, i.e., biochemistry, pharmaceuticals, metallurgy, and medical sciences, where speed, specificity, and accuracy are of paramount importance. Its usefulness for the detection and determination of pesticides and their residues is slowly gaining world-wide recognition, judging from the increasing number of papers appearing in the literature.

The increasing use of the organophosphorus pesticides with their multitude of altered products makes the study of the properties of the altered products and the kinetics of their formation essential to our understanding of these important compounds. In these studies oscillo-polarographic techniques should play an important rôle.

## Summary

Recent advances in polarographic instrumentation and techniques have resulted in increased sensitivity, specificity, and speed in regards to the determination of organic compounds. These advances have created a new interest in the application of polarography for the detection and determination of pesticides and their residues. The polarographic technique which is being utilized most successfully for residue studies is oscillographic polarography.

Oscillographic polarography, its applications to pesticides and their residues, and its advantages and disadvantages, are discussed. A general procedure is presented showing the steps that may be taken in applying polarographic techniques to pesticide residue analysis, including some of the problems that may be encountered and how they were solved. Recent applications of oscillographic polarography to pesticide residue determinations are also discussed.

## Résumé *

Les récents progrès accomplis dans l'instrumentation et les techniques polarographiques ont permis d'accroître la sensibilité, la spécificité et la rapidité des dosages de composés organiques. Ces progrès ont créé un nouveau champ d'intérêt: celui de l'application de la polarographie à la détection et au dosage des pesticides et de leurs résidus. La technique polarographique qui convient le mieux pour les études sur les résidus est la polarographie oscillographique.

Les avantages et les inconvénients de la polarographie oscillographique, de ses applications aux pesticides et à leurs résidus sont discutés. Un procédé général indiquant les stades qui peuvent être franchis an appliquant les techniques polarographiques à l'analyse des résidus de pesticides, et incluant la façon de résoudre certains problèmes qui peuvent se poser, est présentée. Les applications récentes de la polarographie oscillographique aux dosage des résidus de pesticides sont aussi discutés.

## Zusammenfassung **

Die jüngsten apparativen und arbeitstechnischen Fortschritte auf polarographischem Gebiet haben zu einer erhöhten Empfindlichkeit, Spezifität und Schnelligkeit bei der Bestimmung organischer Verbindungen geführt. Damit ist auch ein neues Interesse für den Einsatz der Polarographie zum Nachweis und zur Bestimmung von Pesticiden und ihrer Rückstände erwacht. Für Rückstandsuntersuchungen ist es besonders die oszillographische Polarographie, die mit höchstem Erfolg eingesetzt wird.

Diese Technik, ihre Anwendung auf Pesticide und ihre Rückstände sowie ihre Vor- und Nachteile werden besprochen. An Hand eines allgemeinen Verfahrens wird gezeigt, welche Schritte zur Verwendung der polarographischen Analyse auf Pesticid-Rückstände unternommen werden können, wobei einige der dabei möglicherweise auftretenden Probleme und deren Bewältigung behandelt werden. Darüber hinaus werden neuere Anwendungen der oszillographischen Polarographie auf dem Gebiet der Analyse von Pesticid-Rückständen diskutiert.

## References

AVERELL, P. R., and M. V. NORRIS: Estimation of small amounts of O,O-diethyl O,p-nitrophenyl thiophosphate. Anal. Chem. 20, 753 (1948).

BATES, J. A. R.: Polarographic determination of azinphos-methyl residues in certain crops. Analyst 87, 786 (1962).

* Traduit par S. DORMAL VAN DEN BRUEL.
** Übersetzt von H. FREHSE.

BREZINA, M., and P. ZUMAN: Polarography in medicine, biochemistry, and pharmacy. (Translated by S. Wawzonek) 2 Vols. New York-London: Interscience 1958.

DAVIDEK, J., and G. JANICEK: Polarographic determination of DDT. Experienta 17, 473 (1961).

GAJAN, R. J.: Application of oscillographic polarography to the determination of organic phosphorous pesticide residues. J. Assoc. Official Agr. chemists 45, 401 (1962).

— Application of oscillographic polarography to the determination of organic phosphorous pesticide residues. II. A rapid screening procedure for the determination of parathion in some fruits and vegetables. J. Assoc. Official Agr. Chemists 46, 216 (1963).

GETZ, M. E.: Six phosphate pesticide residues in green leafy vegetables: Cleanup method and paper chromatographic identification. J. Assoc. Official Agr. Chemists 45, 393 (1962).

GIANG, P. A., and R. L. CASWELL: Polarographic determination of O,O-dimethyl-2,2,2-trichloro-1-hydroxyethyl phosphonate. (Bayer L 13-59.) J. Agr. Food Chem. 5, 753 (1957).

HEYROVSKY, J.: The development of polarographic analysis. Analyst 81, 189 (1956 a).

—, and J. FOREJT: Oscillographische Polarographie. Z. Phys. Chem. 193, 77 (1943).

— Anwendung des Kathodenstrahloscillographen in der Polarographie mit Wechselstrom. Chem. Techn. 9, 257 (1957).

— Trends in polarographic analysis. Chemical Age 74, 1449 (1956 b).

JURA, W. H.: Polarographic determination of S-(1,2-dicarbethoxyethyl)-O,O-dimethyl dithiophosphate (malathion). Anal. Chem. 27, 525 (1955).

KEMULA, W.: Chromato-polarographic studies. Roszniki Chem. 26, 281 (1952); Ibid. 26, 694 (1952).

—, and K. KRZENINSKA: Chromato-polarographic studies. Chem. Analit. 1, 29 (1956 a); Ibid. 1, 56 (1956 b).

—, D. SYBILSKA, and J. GEISLER: Chromato-polarographic studies. Chem. Analit. 1, 36 (1956 c).

—, and A. KRZEMINSKA: Chromato-polarographic studies. Chem. Analit. 5, 611 (1960).

MARTENS, P. H., and P. NANGNIOT: La determination de residues d'insecticides et de fongicides par la methode polarographique. Residue Reviews 2, 26 (1963).

— —, and G. DARDENNE: Comparaison de methodes colorimetrique et polarographique en vue du dosage de traces d'insecticides, de fongicides, et d'herbicides organiques nitres. Mededel Landbouwhog. Opzoekst. Gent. 26, 1523 (1961).

MATHESON, L. A., and N. NICOLS: The cathode ray oscillograph applied to the dropping mercury electrode. Trans. Electrochem. Soc. 73, 193 (1938).

MÜLLER, R. H., R. L. GARMAN, M. E. DROZ, and J. PETRAS: The cathode ray polarograph. Ind. Eng. Chem., Anal. ed. 10, 339 (1938).

NANGNIOT, P.: La polarographie applique au dosage de petites doses d'insecticides et de fongicides. I. Le soufre elementaire. Bull. Inst. Agron. Stat. Réch. Gembloux 28, 276 (1960 a).

— Possibilities de dosage de traces de TMTD par la voie polarographique. Mededel Landbouwhog. Opzoekst. Gent. 25, 1285 (1960 b).

— La polarographie applique au dosage de petites doses d'insecticides et de fongicides. II. Le derives de le dithio carbamique. 1. Le Ziram, 2. Le Ferbam, 3. Le Zineb. Bull. Inst. Agron. Stat. Réch. Gembloux 28, 365 (1960 c).

REYNOLDS, G. F., and H. M. DAVIS: An improved Randles type cathode ray polarograph. Analyst 78, 314 (1953).

SANDI, E.: Beiträge zur Analyse einiger Insecticiden Thiophosphorsäureester. Z. Anal. Chem. 167, 241 (1959).

SOHR, H.: Das oszillopolarographische Verhalten einiger organischer Phosphorbindungen. Chem. Zvesti. 16, 316 (1962).

WAWZONEK, S.: Organic polarography. Anal. Chem. 21, 61 (1949); Ibid. 22, 30 (1950); Ibid. 24, 32 (1952); Ibid. 26, 65 (1954); Ibid. 28, 638 (1956); Ibid. 30, 661 (1958); Ibid. 32, 145 R (1960); Ibid. 34, 182 R (1962).

# Polarography for the determination
# of organic feed medicaments

By

Paul T. Allen* and Herman Beckman*

With 2 figures

## Contents

## I. Introduction

Present methods for the analysis of feed medicaments do not ordinarily include polarography but usually consist of colorimetric, ultraviolet, chromatographic or chemical procedures. While many of these methods are well established and quite satisfactory, some have limitations in terms of time required for analysis, their non-specific nature, or erratic results. The use of polarography for the analysis of feed additions has been investigated by several groups, but has not received generally adequate recognition. The purpose of this presentation, then, is to show that polarography has a place in feed additive analysis and may be used in conjunction with other methods such as infrared spectrophotometry or gas chromatography for qualitative and quantitative interpretation of data. There are applications to metabolism studies and other research projects involving feed additives that may lead to improved analytical procedures or ways to study the effectiveness of these compounds.

* Agricultural Toxicology and Residue Research Laboratory, University of California, Davis, California.

## II. Definition of terms

In the context of this paper certain terminology will have meanings of a nature more narrow than might be generally ascribed to the words in common usage. The use of the term *medicament* will infer *feed medicament*. A medicament will then refer to a compound added to the animal feed to produce some desired effect. The desired effect may range through parasite removal, disease prevention, disease control, growth promotion, pigmentation control, transquilization, and perhaps others. Most of the compounds are organic and for the purposes of the discussions of polarography they are synthetic organic materials. Medicaments derived from natural products or inorganic material will not be referred to. A *feed additive* or simply *additive* will be considered equivalent to a medicament in this presentation. Generally, other compounds classed as additives, such as antioxidants, urea, minerals, and vitamins are not considered as medicaments and will not be discussed in this review. Animal feeds of all types are included and they refer to feeds for livestock, poultry, and other small animals. The term *drug* is often used in the literature referring to a feed medicament or feed additive and may be so used in this paper.

## III. Historical (BECKMAN 1959 a)

### a) Use of feed additives

Several attempts to prevent or control diseases and parasites in poultry and domesticated animals with medicated feeds were made as far back as 30 years ago. Colloidal iodine was one of the first medicaments used as an additive in feeds and drinking water for worm removal and for coccidiosis control in poultry, as well as for the control of blackhead in turkeys. Potassium permanganate has been added to poultry drinking water for many years for some nebulous reason. This, no doubt, was an early attempt to ward off disease or to improve general vigor.

The use of finely-ground tobacco or tobacco dust for the control of roundworm infestation in poultry was begun in the early 1930's. The tobacco dust was mixed with the feed as a means of introducing the material into the birds. Nicotine compounds later replaced tobacco dust and were also administered to the birds by means of the feed. Another compound that was found to be effective as an anthelmintic in poultry was sodium fluoride administered with the feed. Sodium fluoride is still used to some extent as a feed medicament for worm control in swine.

Sulfaguanidine was given widespread publicity in the late 1940's for the control of coccidiosis. It was one of the first in a succession of sulfa drugs to gain popularity as a feed additive for disease control. Sulfaquinoxaline, which also saw early use, is still a popular drug for both the preventation and control of outbreaks of coccidiosis.

Within the past ten years the practice of adding various drugs to animal feeds has become widespread. The increase in diversity of nonantibiotic drugs present in medicated feeds has brought a corresponding increase in analytical problems. These problems are especially critical to the regulatory laboratories of both state and federal agencies. MERWIN (1956) emphasized the need for more analytical methods in the following quotation: "Most of the difficulties stem from the lack of sufficient official methods for determining all different drugs, of which about 22 are now used. Although seven authoritative methods are sponsored by the Association of Official Agricultural Chemists, more than three times that many should be available". The seven methods referred to are for arsanilic acid, enheptin, sulfaguanidine, sulfaquinoxaline, sulfanilamide, phenothiazine, and diethylstilbestrol (HORWITZ 1960). In the period since the above statement was made, the list of available feed additives has grown to 60 or more. The 1960 volume of the Methods of Analysis of the Association of Official Agricultural Chemists (A.O.A.C.) lists 15 methods for nonantibiotic drugs. The picture, however, is not as bad as would be inferred from the above citation. There is a number of methods available that are the result of work by several in the feed analysis field. Just because a method has not been subjected to collaborative study by an A.O.A.C. group does not mean that the method is not good. In fact, some methods may be better than some of those labeled as "official".

### b) Analytical methods

**1. Colorimetric.** Table I lists common and chemical names of drugs which have been successfully colorimetrically assayed when present in a feed. There have been methods suggested for several other drugs, 3-nitro-4-hydroxyphenylarsonic acid, $N,N'$-di-(3-nitrobenzene-sulfonyl)ethylenediamine, acetyl-($p$-nitrophenyl)-sulfanilamide, 4-nitrophenylarsonic acid, and 3,5-dinitrobenzamide, which are added to feeds, but experience has shown these methods to be of questionable value (CAVETT 1956 a, b, c, and d; CAVETT and HEOTIS 1958 and 1959). Several causes for failure of these methods may be cited. First, other drugs that may be present may interfere by simultaneously reacting to give rise to colored products. Second, a colored material already present in the feed may be extracted along with the drug to be determined and resulting in high spectrophotometer readings. Some of the colored materials that might be extracted are carotenoids, chlorophylls, anthocyanins, and highly-colored partially-soluble proteinaceous compounds. Overheating or biological oxidation of the feed may also give rise to colored extractable products. Third, the wrong solvent has been suggested for use in certain cases. Fourth, the drug may form an insoluble complex with protein, etc., which may cause the method to fail.

Several examples of determinations that fail because of the above enumerated causes may be cited. In the colorimetric method for the determination of 3-nitro-4-hydroxyphenylarsonic acid (CAVETT 1956 c), the extraction procedure also extracts pigments from the feed which, in turn, obscure the colored product sought. The same method is also subject to interference by other drugs, especially acetyl-($p$-nitrophenyl)sulfonilamide and $N,N'$-di(3-nitrobenzenesulfonyl)ethylenediamine. In the determination of dienestrol diacetate (LIMECREST 1956), it has been observed that the solvent specified for use is ineffective.

Workers at the Connecticut Agricultural Experiment Station (MADER 1956) found prolonged alkaline hydrolysis of feeds liberated a material

Table I. *Feed additives that have been colorimetrically assayed*

| Common name | Chemical name |
|---|---|
| Acetyl Enheptin . . . . . . . | 2-acetyl-amino-5-nitrothiazole |
| Amprolium . . . . . . . . . | 1-(4-amino-2-*n*-propyl-5-pyrimidinyl-methyl)-2-picolinium chloride HCl |
| APNPS . . . . . . . . . | acetyl-(*p*-nitrophenyl)sulfanilamide |
| Arsanilic acid . . . . . . . | *p*-aminobenzenearsonic acid |
| Arsenosobenzene . . . . . . | phenylarsine oxide |
| Enheptin . . . . . . . . . | 2-amino-5-nitrothiazole |
| Bithionol . . . . . . . . . | 2,2'-dihydroxy-3,3'5,5'-tetrachlorodiphenyl sulfide |
| Cadmium anthranilate . . . . | — |
| Cadmium oxide . . . . . . | — |
| Carbarsone . . . . . . . | *p*-ureidobenzenearsonic acid |
| Chlortetracycline . . . . . . | aureomycin |
| Di-*n*-butyltin dilaurate . . . . | — |
| Dienestrol diacetate . . . . . | 3,4-bis(*p*-hydroxyphenyl)-2,4-hexadienediacetate |
| Diethylcarbamazine . . . . . | 1-diethyl carbamyl-4-methyl-piperazine |
| Diethylstilbestrol . . . . . . | 3,4-bis(*p*-hydroxyphenyl)-3-hexene |
| Dimetridazole . . . . . . . . | 1,2,5-nitroimidazole |
| Dinsed . . . . . . . . . . | dinitrodiphenylsulfonylethylenediamine |
| DNBA . . . . . . . . . . | 3,5-dinitrobenzamide |
| Dynafac . . . . . . . . . . | trimethyloctadicyl and hexadecyl ammonium stearate M. W. 590 |
| Furazolidone . . . . . . . | *N*-(5-nitro-2-furfurilidene)-3-amino-2-oxazolidone |
| Glycarbylamide . . . . . . . | 4,5-imidazole-dicarboxamide |
| Hitchings antimalarial compound . . . . . . . | 2,4-diamino-5-(*p*-chlorophenyl)-6-ethylpyrimidine |
| Methiotriazamine . . . . . . | 4,6-diamino-1-(4-methylmercaptophenyl)1,2-di-hydro-2,2-dimethyl-1,3,5-triazine |
| Nicarbazine . . . . . . . . | 4,4'-dinitrocarbanilide complex with 2-hydroxy-4,6-dimethyl pyrimidine |
| Nicotine . . . . . . . . . | 1-methyl-2-(3-pyridyl) pyrolidine |
| Ninhydrazone (HC-067) . . . | — |
| Nithiazide . . . . . . . . | 1-ethyl-3-(5-nitro-2-thiazone) urea |
| Nitrofurazone . . . . . . . . | 5-nitro-2-furaldehyde semicarbazone |
| | 3-nitro-4-hydroxyphenylarsonic acid |
| Nitrophenide . . . . . . . | bis(*m*-nitrophenyl)disulfide |
| | 4-nitrophenylarsonic acid |
| PABA . . . . . . . . . . | *p*-aminobenzoic acid |
| Penicillin . . . . . . . . . | benzylpenicillinic acid |
| Perphenazine . . . . . . . | 1-(2-hydroxyethyl)-4-[3-(2-chloro-10-phenothiazinyl)-propyl]-piperazine |
| Phenothiazine . . . . . . . | thiodephenylamine |
| Piperazine . . . . . . . . | diethylenediamine |
| Promazine . . . . . . . . | 10-(3-dimethylaminopropyl) phenothiazine |
| Ronnel . . . . . . . . . . | O,Odimethyl-O-2,4,5-trichlorophenyl phosphorothioate |
| Streptomycin . . . . . . . | — |
| Sulfaquinoxaline . . . . . . . | 2-sulfanilamidoquinoxaline |
| Zoalene . . . . . . . . . | 3,5-dinitro-*o*-toluamide |

which showed a tendency to produce an interfering color when the feed was analyzed for the presence of arsanilic acid. After the interfering material was identified as tryptophan, means were found to avoid its liberation during analysis. This was done by preventing excessive hydrolysis through control of pH and by limiting the strength of the diazotizing reagents.

Drugs which are added to feeds at the present time cover the range from strictly inorganic compounds to rather complex heterocyclic compounds and antibiotic-type materials. Thus, there can be no general type of method for analysis. Frequently, colorimetric determinations are used, although titrimetric and gravimetric methods may also be employed. The greatest interest in the drug-analysis field has been directed toward colorimetric procedures. This interest is in part due to the sensitivity that may often be achieved with a colorimetric procedure and the availability of simple, low-cost colorimeters and spectrophotometers. In most cases, anyone can be trained to use this equipment. With colorimetric methods, then, the limiting phase of the analytical procedure is the handling of the sample up to the point of reading the developed color.

The following five drugs were previously studied by one of the present authors to develop colorimetric analyses specifically for use with mixed feeds (BECKMAN 1958, BECKMAN and DeMOTTIER 1959, BECKMAN and FELDMAN 1959 and 1960).

*Furazolidone* and *Nitrofurazone*. Studies to determine biological effectiveness of various nitrofuran compounds were conducted by the Norwich Pharmacal Company beginning in the late 1940's, and tentative procedures for the analysis of furazolidone and nitrofurazone were released (NORWICH 1950) in May 1950.

Furazolidone may be used in feeds in concentrations ranging from 0.00083 percent to 0.022 percent for one or more of the following purposes:

Poultry
1. Aid in growth promotion
2. Improve feed efficiency
3. Stimulate egg production
4. Improve feed-egg ratio
5. Prevent fowl typhoid pullorum and paratyphoids
6. Prevent paracoloron and coccidiosis
7. Treat histomoniasis (blackhead)
8. Treat infectious arthritis

Rabbits
1. Prevent enteritis (Mucoid and Diarrhea)
2. Prevent Pasteurella-type pneumonia

Nitrofurazone may be used in feeds in concentrations ranging from 0.0056 percent to 0.056 percent for one or more of the following purposes:

1. Prevent coccidiosis outbreaks in poultry
2. Aid in growth promotion, improve feed efficiency and amprove pigmentation in poultry
3. Prevent mortality from pullorum in poultry
4. Treat bacterial swine enteritis caused by *Salmonella koleraesius*

A mixture of furazolidone (0.00083 percent) and nitrofurazone (0.0056 percent) is also marketed as a feed additive under the trade name Bifuran.

Several methods for the analysis of furazolidone and nitrofurazone hav been suggested and used. BUZARD *et al.* (1956) utilized the product formed by the reaction of nitrofurazone and/or furazolidone with phenylhydrazine. The phenylhydrazone is extracted from the reaction mixture with toluene, and its absorbance measured with a spectrophotometer. The procedure gives good results with feeds containing not less than 0.0055 percent of either, or a combination of the two drugs, but is quite lengthy. The procedure does not differentiate between the two drugs, since both compounds yield the same phenylhydrazone.

ELLS et al. (1953) described the following general procedure. The drug furazolidone, as well as nitrofurazone, is extracted from the feed, with ethyl alcohol and the extract divided into two portions. One portion is treated with sodium hydrosulfite to reduce the drug. This reduced solution is used as the blank for the direct spectrophotometric measurement of the unreduced portion of the extract.

NAGASAWA and OHKUMA (1954) proposed a procedure using 2,4-dinitrophenylhydrazine to form insoluble dinitrophenylhydrazone derivatives. These derivatives are filtered from the solution, and the assay is completed as a gravimetric procedure.

The method of HESS and CLARK (1954) is essentially the procedure of ELLS et al. (1953).

None of the above methods were designed for analyzing samples containing less than 0.0055 percent of either drug. They all leave some feature to be desired, in three cases (ELLS 1953, NAGASAWA and OHKUMA 1954, HESS and CLARK 1954) reproducibility, and in one case (BUZARD et al. 1956) speed.

A recent paper by BRÜGGEMANN et al. (1962) describes the analysis of nitrofurazone in poultry feeds. A comparison of two methods (BUZARD et al. 1956; BECKMAN 1958) is made and evaluations summarized on the efficiency and reprocibility of the two methods. Improved extraction and cleanup procedures are described for the recovery of this compound. It may be of interest, also, to point out that the paper by BECKMAN (1958) describes the only procedure for separating furazolidone and nitrofurazone allowing each compound to be determined separately.

CROSS et al. (1960) described an extraction and colorimetric procedure for the analysis of nitrofurazone in feeds. The paper presented nothing new from the standpoint of color development, sensitivity, or specificity, but does have a rather extensive literature review.

*3,5-Dinitrobenzamide.* The compound 3,5-dinitrobenzamide (DNBA) is now being marketed as a medicinal additive for poultry feeds. It is used as an aid in the prevention or treatment of fowl typhoid, paratyphoid, and pullorum disease, and as an aid in stimulating growth in poultry flocks. The drug may be found in feeds ranging from 0.025 percent to 0.15 percent, and may be mixed with one of the following:

1. Arsanilic acid
2. Sodium arsanilate
3. 3-Nitro-4-hydroxyphenylarsonic acid
4. Furazolidone
5. Furazolidone, 3-nitro-4-hydroxyphenylarsonic acid, and nitrofurazone
6. $N^4$-Acetyl-$N^1$(4-nitrophenyl)sulfanilamide and 3-nitro-4-hydroxyphenylarsonic acid

JANOVSKY and ERB (1886) investigated the potassium hydroxide cata-
lyzed reactions of various bromo- and nitro-derivatives of azobenzenes
with acetone. The products of these reactions were colored. JANOVSKY (1891)
extended this work to include a study of dinitro derivatives of various
phenols, chlorobenzenes, and alkylbenzenes. A variety of colors from red
to blue were observed. A year later, VON BITTO' (1892) reacted various
aldehydes and ketones, other than acetone, with metadinitroaromatics to
determine if colored products would be formed. REITZENSTEIN and STAMM
(1910) investigated the mechanism of the base-catalyzed reaction of acetone
with metadinitrochlorobenzene.

RUDOLPH (1921) showed that certain di- and trinitro compounds would
give various colors (purple to red) with acetone, using ammonium hydroxide
as the base. He also investigated the base-catalyzed reaction of the nitro
compounds with alcohol. The dinitro compounds investigated were o, m,
and p-dinitrobenzene, 2,4- and 2,6-dinitrotoluene, and 1,8-dinitronaphtha-
lene. Only 2,4-dinitrotoluene produced a color (blue) in alcohol with sodium
hydroxide, and only 1,8-dinitronaphthalene produced a color (red) in alcohol
with ammonium hydroxide. No color was produced in any case by o-
dinitrobenzene or by 2,6-dinitrotoluene. BOST and NICHOLSON (1935) proposed a method for distinguishing among mono-, di-, and trinitro compounds using the JANOVSKY (1891) reaction. Although they noted that
certain compounds did not give the expected color, the following generali-
zations were made: mononitro compounds form no color, dinitro compounds develop a violet color, and trinitro compounds develop a red color.
The test was performed by dissolving 0.1-g. samples in a mixture of 10 ml.
of acetone and 3 ml. of five percent aqueous sodium hydroxide solution.
Sensitivity for the test was claimed to be one part in 1.5 million, or
0.67 parts per million (p.p.m.).

ENGLISH (1948) introduced a new non-aqueous alkali reagent for color
development. He dissolved sodium monoxide ($Na_2O$) in alcohol and diluted
the resulting solution with acetone. It was noted that substitution of ammonia for the sodium monoxide was ineffective, and that the substitution of
potassium hydroxide gave colors that were too fugitive for use. ENGLISH
(1948) also suggested that the color formation is due to quinoidation. This
cannot be the case, as meta-substituents on a benzene ring cannot enter
into a quinoidal configuration.

BAERSTEIN (1943) utilized butanone and an alkali to produce colors
with nitro derivatives of benzene and toluene. He studied the effect of
alkali concentration on color intensity and postulated that the mechanism
of the reaction involved the enol-form of the ketone and the aci-form of
one of the nitro groups. At least one para-position must be free for the
reaction to proceed. This reaction is similar to the mechanism suggested by
REITZENSTEIN and STAMM (1910). It would seem likely that there would
be involvement of both nitro groups simultaneously, and that, in such a case,
no quinone structure could be postulated. The carbonyl group is not essential to the reaction, as shown by RUDOLPH's (1921) investigation with
alcohol. A meta-dinitro compound, as 2,6-dinitrotoluene, fails to respond to
the color test, even though both para-positions are free. Evidently a

requirement for color production is that the position between the nitro groups be open. This might well be explained on the basis of steric hindrance. It also gives evidence of the participation of both nitro groups in the color reaction. It has been further shown in this laboratory that a color can be produced by certain dinitro compounds, utilizing liquid anhydrous ammonia as both the solvent and alkali source.

The JANOVSKY (1891) reaction was also utilized by FISCHER (1950) for the macroanalysis of various polynitro compounds. He used 20 to 50 percent potassium hydroxide solutions in this method.

The mechanism of the color formation in the JANOVSKY (1891) reaction was recently studied by FOSTER and MACKIE (1962). Comparisons were drawn with the well-known ZIMMERMANN reaction and postulates were made to describe the interactions of the ketone, the alkali, and the meta-dinitrobenzenes. The source of the alkali and the amount of ketone have a direct bearing on the type and amount of color formed.

A variety of other methods for the determination of other specific compounds, mainly dinitrophenols, have been published. These, however, do not fall within the scope of the colorimetric reaction being studied here. They include titrimetric, gravimetric, and colorimetric methods.

None of the above workers have attempted to obtain color production with any dinitrobenzamides (DNBA).

*Cadmium anthranilate.* Cadmium anthranilate, as a feed additive, is intended for use solely as an anthelmintic for poultry or swine. It must be used alone in a prepared feed and only at 0.055-percent level.

Several methods are available in the literature which may be utilized for the analysis of cadmium anthranilate. These, however, depend on the analysis of the cadmium portion. There is no published method designed specifically for the analysis of cadmium anthranilate.

An electrolytic deposition method for cadmium has been described by SCOTT (1939) and adapted for feeds and concentrates by STONE (1955). The method entails complete digestion of the feed with nitric, sulfuric, and perchloric acids. The solution is then treated in the usual manner for electroplating onto a platinum gauze electrode.

A second method by STONE (1956) utilizes the same preparative technique, but the cadmium is determined with dithizone. A previous report of this method was made by KLEIN and WICHMANN (1945), who investigated cadmium as a contaminant in foods. The dithizone method for cadmium is described in detail by SALTZMANN (1953), who also gives instructions for removing interfering ions. Perhaps the most precise method for determining cadmium is the polarographic technique described by KOLTHOFF and LINGANE (1952). All of the above methods involve the complete digestion of the sample and of the cadmium anthranilate in order to obtain a solution of cadmium as a salt. In the polarographic method a direct ashing procedure may be utilized. A dithizone extraction is used to removing interfering cations.

*Piperazine.* Piperazine has been used as a medical agent for the treatment of gout and rheumatism, and more recently as an anthelmintic in humans and animals. A large array of other uses, industrial and medicinal, has been developed and suggested for piperazine derivatives (*Dow Chemical Co.* 1956, *Jefferson Chemical Co.* 1956). Low-level feeding of piperazine phosphate (20 g./100 lbs.) to swine has been reported to cause an improvement in the growth rate. The increased growth rate is due to an increase in feed consumption, and is not necessarily the result of the worm-killing effect of piperazine. Piperazine has been shown to be relatively innocuous, especially toward animals capable of emesis. The product is generally marketed as an inorganic acid salt for addition to feed formulations, but may also be obtained as the anhydrous or hydrated compound or as a salt of an organic acid. A large amount of work has been published concerning piperazine, as evidenced, for example, by the two reviews cited above. Several of the references cited in these reviews concern methods of analysis. However, none of these has been used for the analysis of piperazine in animal feed.

A recent publication by LENG (1957) describes an analytical method for determining piperazine in feeds. Piperazine salts are converted to the free base by the use of an alcoholic alkali solution which avoids appreciable breakdown of fat or protein. The extract is acidified and passed through an ion-exchange resin column. The piperazine is eluted from the resin and converted to the dipicrate for gravimetric determination. The method requires rather large samples, and is too time-consuming for routine use.

Several semimicro methods for the analysis of piperazine have been published, but they all lack the sensitivity required for parts per million analysis (BARNHARD 1947, CASTIGLIONI 1939 and 1940, VITTE and COUSTOU 1943). The use of chloranil (CRIPPA 1953), tetrachloro-1,4-benzoquinone, and *p*-quinone (FOUCRY 1934) for the identification of the presence of certain amines have been reported. Chloranil is used to produce colors on paper chromatograms to indicate the presence of phenols, naphthylamines and aniline derivatives. Quinone is reported to react in an acid or neutral medium to detect the presence of a variety of amines.

FUSON (1950), in a discussion of the reactions between carbonyl compounds and amines, states that, in addition to the carbonyl compound and the amine, a base is necessary. In the case of *p*-quinone and piperazine, the quinoidation is enhanced by the alkaline medium, and a colored complex is formed. The reaction of a quinone with an amine is also analogous to that described by FEINSTEIN (1952) for pyrethrins and allethrin. The reaction between the carbonyl group of these insecticides and an alkaline solution of 2-(2-aminoethylamine)-ethanol produces a colored product in the presence of elemental sulfur. FEINSTEIN (1952) also found that other alkalies and other amines produce colors and, under certain conditions, certain other carbonyl compounds will respond to the color test without the use of sulfur.

**2. Chemical.** Of the animal feed medicaments in common use, the principal compounds analyzed by chemical methods are parts of those containing arsenic and cadmium. The review of the cadmium methods was previously given under the colorimetric section above. Arsenic has been the subject of probably hundreds of papers. However, two of these have been applied to feeds and are considered quite reliable (HOFFMAN 1956, *Fisher Scientific Co.* 1960). Although this is listed under chemical methods, and indeed most of the procedure requires considerable dexterity in handling chemical procedures, the final measurement is colorimetric. It should be pointed out that in the molybdate blue method, the absorbance maximum is at or near 850 m$\mu$ instead of 750 m$\mu$ as is often specified. A ratio recording double beam spectrophotometer will verify this and show that at 750 m$\mu$ the reading is taken on a steep slope of the curve. The lower wavelength was probably developed with the use of less expensive colorimeters that would not read beyond 750 m$\mu$.

3. **Instrumental: gas chromatography and infrared.** *Gas chromatography.* Instrumentation that has been applied to the analysis of animal feed medicaments other than colorimetric and ultraviolet includes infrared, gas chromatography, and polarography. In a recent paper (BECKMAN and ALLEN 1962) the authors described the application of gas chromatography and infrared spectroscopy to the analysis of a number of feed additives. In this paper it was pointed out that a number of the commonly used medicaments are readily resolved by gas chromatography. Table II shows a list

Table II. *Feed additives that have been chromatographed by gas-liquid chromatography*

| Common name | Chemical name |
|---|---|
| Bithionol . . . . . . . . . . | 2,2'-dihydroxy-3,3',5,5'-tetrachlorophenylsulfide |
| Diethylcarbamazine . . . . . | 1-diethylcarbamyl-4-methyl-piperazine |
| Diethylstilbestrol . . . . . . | 3,4-bis($p$-hydroxyphenyl)-3-hexene |
| DNBA . . . . . . . . . . . | 3,5-dinitrobenzamide |
| Nitrofurazone . . . . . . . . | 5-nitro-2-furaldehyde semicarbazone |
| Nitrophenide . . . . . . . . | bis($m$-nitrophenyl)-disulfide |
| Phenothiazine . . . . . . . . | thiodiphenylamine |
| Piperazine . . . . . . . . . | diethylenediamine |
| Sulfaquinoxaline . . . . . . . | $N'$-(2-quinoxalinyl)-sulfanilamide |
| Zoalene . . . . . . . . . . | 3,5-dinitro-$o$-toluamide |

of those compounds known to show a definite response. For these results a gas chromatograph with a thermal conductivity detector and programmed temperature was used. Normally, as little as 5 $\mu$g. will show a response, but occasionally some compounds require as much as 25 $\mu$g. to show a reasonable signal. Several columns packed with Chromosorb and coated with various silicones from 5 to 20 percent were used. All silicone materials were stable to at least 250° C.

Since this technique can be used for separating components in a mixture and collecting the desired component, this approach to a simplified cleanup has been investigated. The desirability of this type of cleanup followed by polarographic analysis would be of the greatest value for feed additive mixtures, where one compound interferes with the analysis of another. This procedure would also aid in removing interfering feed extractives when measuring the compound by ultraviolet absorption, chemical, or color. In addition to being used as a cleanup procedure, the gas chromatograph can be applied to measurements of trace amounts of metabolites, or as a possible analytical procedure where no method exists for a given feed additive.

In our study of this technique, known quantities of 3,5-DNBA, Zoalene, and acetyl enheptin were injected onto a gas chromatographic column and collected in a Pasteur pipette containing a glass wool plug saturated with, 95 percent ethanol. The exit port of the gas chromatograph was fitted with a rubber septum with a small hole in the center. Before the compounds were eluted, the collector was inserted into the exit port and removed when the material was completely eluted. The collected material was eluted from the collector with 95 percent ethanol into a volumetric flask and subjected to UV analysis. It was found that recoveries were consistently low; in the order of 55 to 60 percent. Recovery of the feed additive from the gas chromatograph was also checked by the polarographic method and found to be in good agreement with the ultraviolet analysis, that is, 60 percent. The halfwave potential of the recovered material was identical to the standard. The fraction collected from the gas chromatograph was subjected to infrared analysis, compared to the spectrum of the original material, and found to be identical. To eliminate the loss from the filaments, a known quantity of material was injected on the gas chromatographic column with the detector off. The recorder chart paper was rerolled to the origin where a standard had been injected while the detector was on. The fraction was collected that corresponded to the original response and the material subjected to ultraviolet analysis. Recovery was increased to 95 percent. Apparently the loss occurred when the compound passed over the hot filaments.

There are a number of additives that will not pass through the gas chromatograph or do so with evidence of degradation or decomposition. For example, the group of organic arsenicals including arsanilic acid, arsenosobenzene, carbasone, 4-nitrophenylarsonic acid, sodium arsanilate, and 3-nitro-4-hydroxyphenylarsonic acid do not chromatograph because of their melting points and low vapor pressures. Another group of medicaments whose vapor pressures are so low that they are not suited for normal gas chromatographic procedures include:

APNPS
Amprolium
Furazolidone
Glycarbylamide
Nicarbazine
Nitrofurazone
Enheptin

It may be of interest to note that there is still hope for the chromatography of these compounds that are so non-volatile that they will not respond at 300° C. Two routes seem to be open. One is in connection with the arsonic acids which would involve the preparation of the ester of the acid which should render the compound more volatile. This process is routinely done with common organic acids and also renders phosphoric acid more volatile. Organic esters of phosphoric acid are common among the pesticides that are known to chromatograph. Since arsenic is in the same family in the periodic table with phosphorus and nitrogen, volatile esters are a distinct possibility.

The second route involves a more drastic step, but could yield very reliable data also. There is now available a new device that can be added to a gas chromatograph, at the injection port, called a pyrolysis unit. It has been determined that when a complex or non-volatile compound is subjected

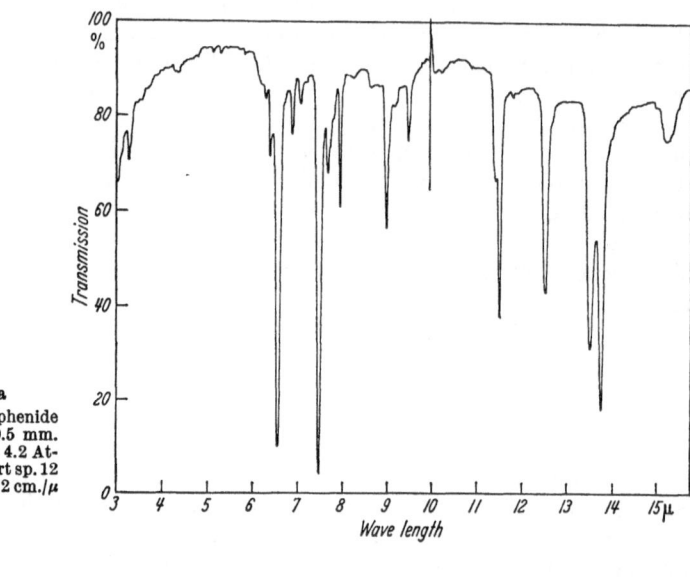

Fig. 1 a
11 μg. Nitrophenide
KBr 1.5 × 0.5 mm.
Slit 970 Gain 4.2 Atten. 1100 Chart sp. 12
Ord. IX Abs. 2 cm./μ

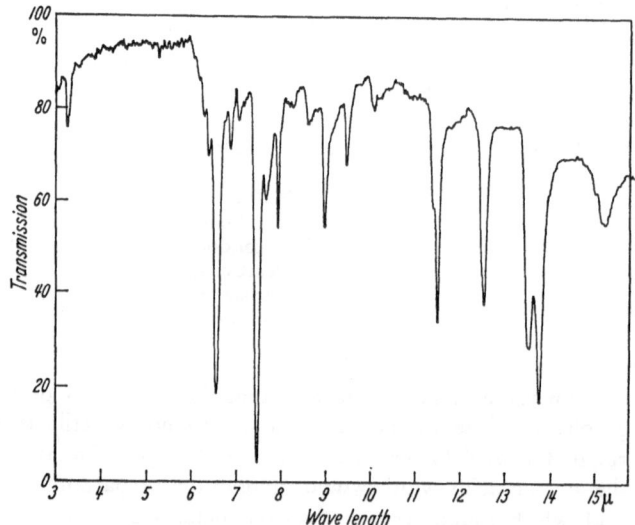

Fig. 1 b
From GLC

to high heat in an inert atmosphere, that it will decompose in a predictable and reproducible fashion. The device contains a platinum coil onto which the sample is deposited. After evaporation of the solvent with mild heat, the coil is inserted into the chromatograph, locked into position, and then heated very strongly and rapidly. When current is applied it attains 1,300° C. in one second and a sample is completely pyrolyzed in three seconds.

After pyrolysis the volatile products are swept into the chromatographic column with the carrier gas and detected as they emerge. For example, if

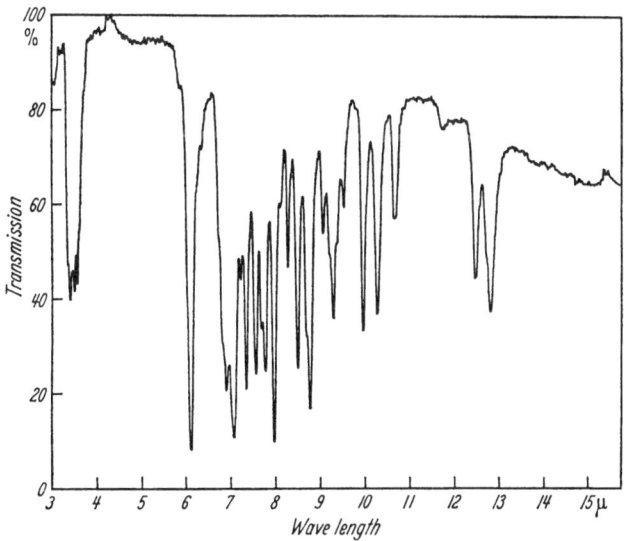

Fig. 1 c
12 μg.
Diethylcarbamazine
KBr 1.5 × 0.5 mm.
Slit 960 Gain 4.3 Atten. 1100 Chart sp. 12
Ord. IX Abs. 2 cm./μ

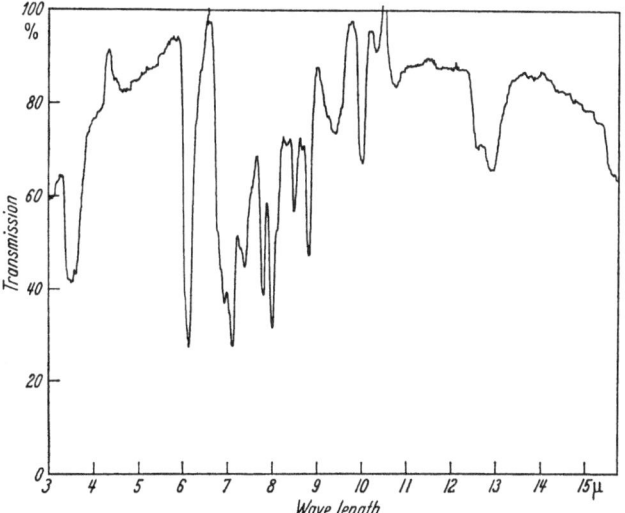

Fig. 1 d
From GLC

3-nitro-4-hydroxyphenylarsonic acid were pyrolyzed, one would expect to see o-nitrophenol, nitrobenzene, phenol, and benzene. It is very doubtful that each of the four expected components would be obtained in equal amounts. One would probably predominate. The unique feature is that the relative peak heights for each of the components is a constant and would be observed with each analysis. This is another means of obtaining both identification and quantitative analysis. This procedure could be applied to those compounds that are not volatile at 300° C.

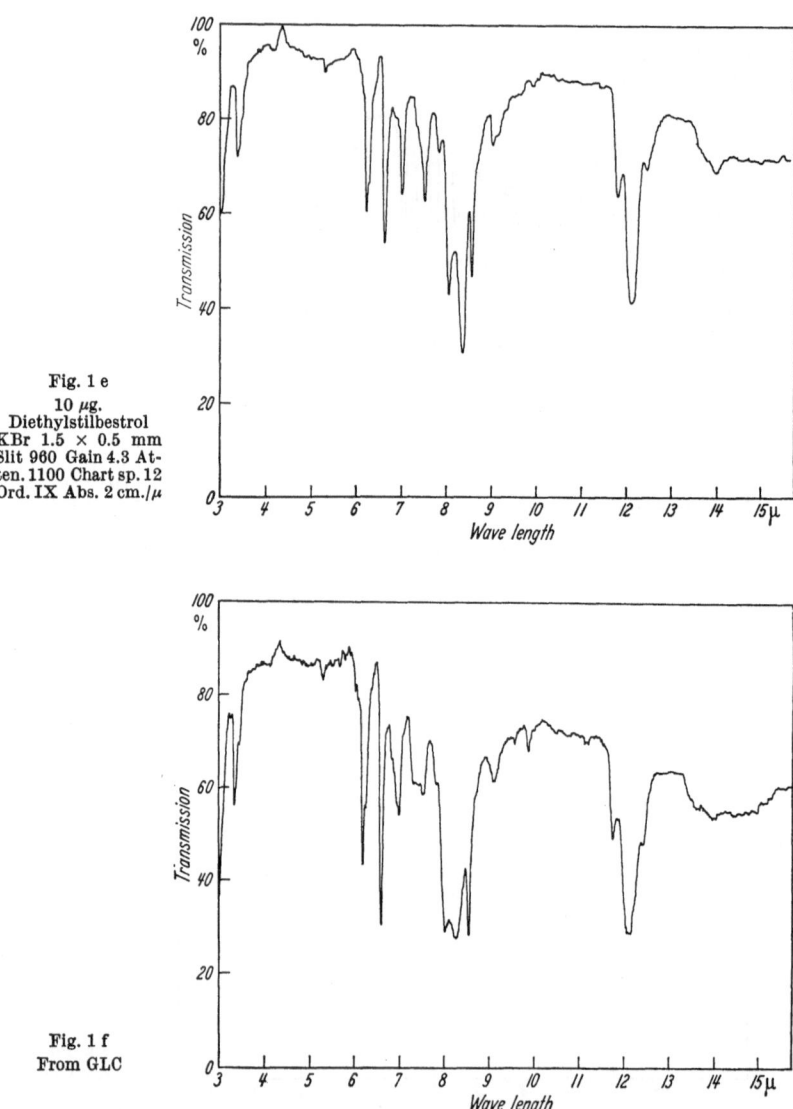

Fig. 1 e
10 μg.
Diethylstilbestrol
KBr 1.5 × 0.5 mm
Slit 960 Gain 4.3 Atten. 1100 Chart sp. 12
Ord. IX Abs. 2 cm./μ

Fig. 1 f
From GLC

Fig. 1. Infrared spectra of nitrophenide, diethylcarbamazine, and diethylstilbestrol recovered from a gas chromatograph with standard spectra shown for comparison

*Infrared.* Of those medicaments that were shown to pass through the gas chromatograph, a number were collected from the effluent gas stream and subjected to analysis by the infrared spectrophotometer. Examination of the spectra showed that the compounds passed through the chromatographic system intact. Some examples of those collections are shown in Fig. 1. Spectra of the reagent grade reference standards are also shown and are presented for comparison purposes. The quantities used (about 10 $\mu$g.)

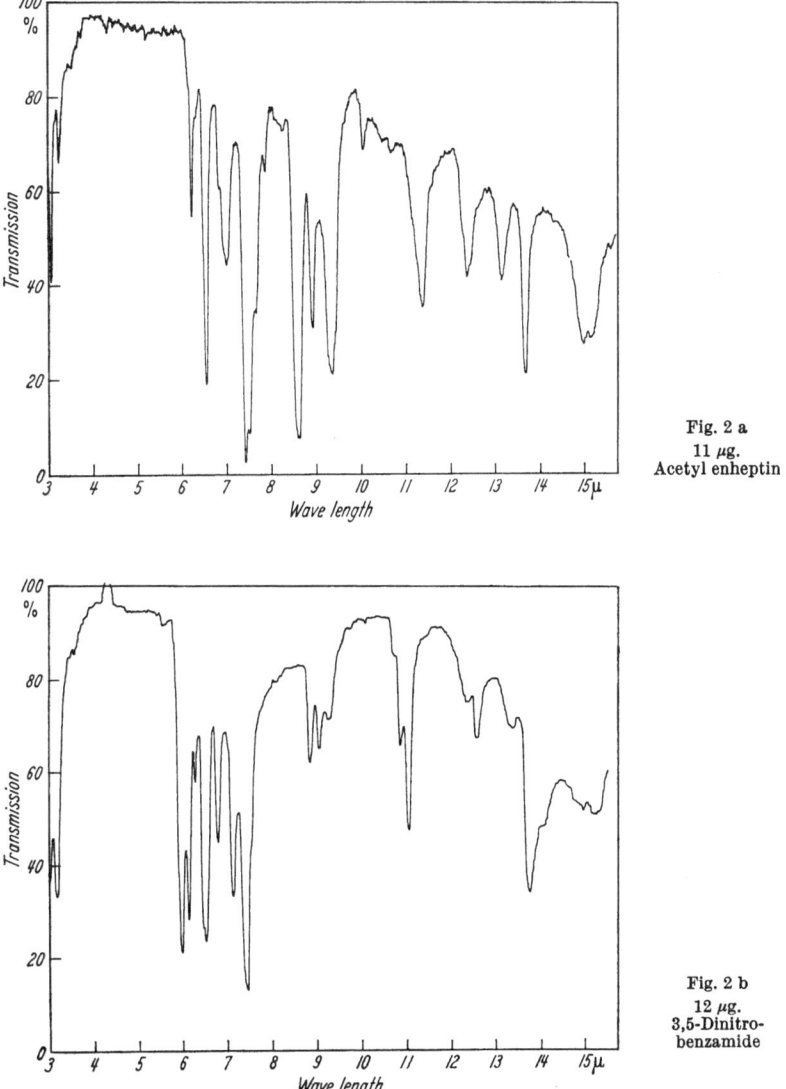

Fig. 2 a
11 $\mu$g.
Acetyl enheptin

Fig. 2 b
12 $\mu$g.
3,5-Dinitro-
benzamide

are representative of the resolution obtainable using micro-potassium bromide discs with a beam condenser.

Infrared spectra of most of the drugs discussed in this paper have been made using the micro-method. These are part of a master reference file for agricultural chemicals in this laboratory. This file includes insecticides, fungicides, herbicides, nematocides, plant growth regulators, and animal feed additives and is coded for ready reference.

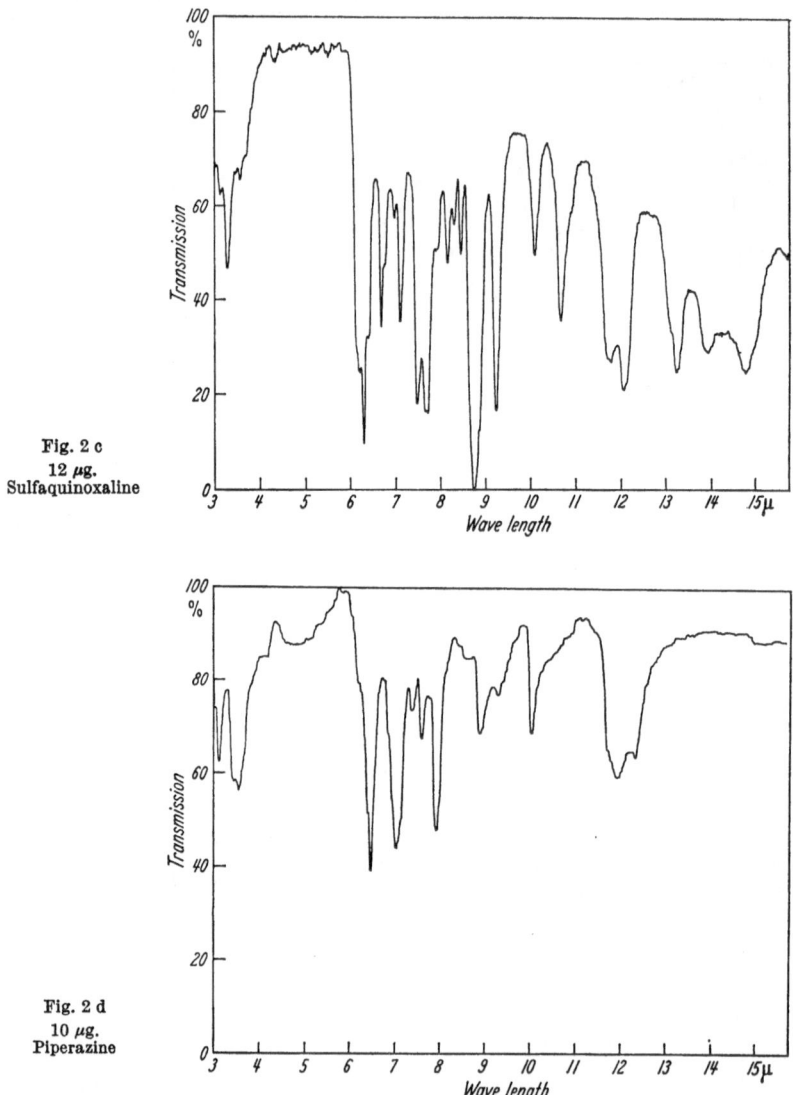

Fig. 2 c
12 μg.
Sulfaquinoxaline

Fig. 2 d
10 μg.
Piperazine

Fig. 2. Infrared spectra of four feed medicament standards in common usage

Several other spectra representative of several chemical types are shown in Fig. 2. These infrared spectra are analogous to fingerprints and are often used for identification purposes. Quantitative analyses have also been made using infrared techniques, even in the parts-per-million range. The conditions of instrument operation are those of Fig. 1a.

**4. Instrumental: Other chromatographic techniques.** With colorimetric methods receiving the greatest attention in terms of new analytical methods during recent years, techniques such as paper chromatography and thin-layer chromatography have been largely bypassed. CASTIGLIONI and NIVOLI (1953) described a paper chromatographic procedure for isolating and identifying piperazine. Chloranil has been used (CRIPPA 1953) as a chromogenic reagent in paper chromatography of piperazine. Quinone was also described as a color reagent for amines (FOUCRY 1934) and would likely be usable as a chromogenic agent in either paper or thin-layer chromatography. CAMPBELL *et al.* (1943) have described the use of *p*-methoxyphenylisocyanate as a reagent for the identification of amines. Probably, there are other reagents of potential value for use as chromogenic agents for either paper or thin-layer chromatography.

Paper chromatography or thin-layer chromatography has been used extensively in the pesticide residue analysis field and could be so applied to feed additives. The techniques are of value for limited quantitative assay, but are of great value for qualitative analysis. These techniques may be used in conjunction with other methods of analysis as a cleanup procedure. An example of this is found (NASH *et al.* 1963) in the technique for the recovery of rotenone from thin layer strips for infrared analysis. The technique has been applied to the recovery of specific naturally occurring plant constituents such as ferulic acid in this laboratory. This technique could be applied equally well to the isolation of a feed additive or for separation of a mixture of medicaments. Drugs recovered from thin layer strips could then be analyzed in a number of ways including polarography, gas chromatography, or infrared spectroscopy.

## IV. Instrumentation for polarography

The use of polarography for the analysis of feed additives has been investigated by several groups. MOORE and GUERTAL (1960) described the polarographic determination of nitrofurans in feed. DAFTSIOS and SCHALL (1962 a), using A. C. polarography, reported on a number of nitro-group-containing feed additives. SASAKI (1958) showed that the diffusion current for nitrofurans was constant between the pH range 4 to 12. One of the purposes of this paper is to show that polarography is broadly applicable and that there are several ways in which it can fit into an analytical program.

One of the characteristics of a reducible compound is its half-wave potential. The diffusion current is also a characteristic of the compound and its concentration. It should be noted that concentrations as low as $10^{-5}$ molar are well within the range of a standard D. C. polarographic instrument. Greater sensitivity is offered with A. C. polarography. A. C. polarographic equipment can be designed from existing D. C. instruments (MILLER 1956 and 1957). A. C. polarography frequently offers greater resolution as well as sensitivity as favorable features. Feed additives are usually found to be in the 10 to 100 p.p.m. range, which is well above the lower limit of D. C. polarography.

When a compound is subjected to polarographic analysis, only a small part of it is reduced and following polarographic analysis the material may be recovered for other uses. Polarography may thus serve as a non-destructive step in a series of qualitative and quantitative sources of information.

In the analysis of feed medicaments by colorimetric or UV means, some type of cleanup is usually necessary, with adsorption column chromatography frequently employed. The polarographic detection of these materials following the usual cleanup practices offers advantages. In many cases an electrolyte may be added before or after chromatography, and the polarogram recorded in the chromatographic solvent.

Through combination of analytical preparation steps, time is saved. The time for analysis must be considered in many laboratories.

A. C. polarographic methods have increased the possibilities of polarography, especially for detecting half-wave potentials as close as 0.05 volt. The A. C. polarogram is essentially the first derivative of the D. C. polarogram. With compounds such as Zoalene (3,5-dinitro-o-toluamide) and 3,5-dinitrobenzamide, each of which give two waves very close together, it is now possible to separate them into sharp peaks by A. C. polarography. The height of the A. C. polarogram peak is proportional to concentration. Daftsios and Schall (1962 b) have also utilized the A. C. polarograph to analyze poultry feeds for 3,5-dinitrobenzamide and acetyl-p-nitrophenyl sulfonamide simultaneously. This method should require less time than the separate colorimetric analysis for the two drugs and should be applicable to other feed additives using either A. C. or D. C. polarography. The simultaneous determination of compounds is possible by polarography if their half-wave potentials are not the same. By using A. C. polarography, where resolution is improved, compounds having very close half-wave potentials can be identified and a quantitative measure of the amount present can be made. The simultaneous determination of two or more compounds by polarography should require less time than the separate colorimetric analysis for the drugs.

Oscillographic polarography is also being used more extensively. For permanent record of analysis, a photograph must be taken of the oscilloscope pattern. This is offset in many cases by the fast time for analysis.

Aromatic nitro compounds are reduced at the dropping mercury electrode and generally give two waves in acid solution. The first wave corresponds to a four-electron reduction of the nitro groups to the corresponding hydroxylamine. The second wave is the reduction of the hydroxylamine to the amine. This wave usually occurs at more negative potential. The reduction can be represented by the following equations (1) and (2). The reduction of an aldehyde is shown as a one-step, two-electron change (3).

$$R - NO_2 + 4e + 4H^+ \rightarrow R - NHOH + H_2O \qquad (1)$$

$$R - NHOH + 2e + 2H^+ + \rightarrow R - NH_2 + H_2O \qquad (2)$$

$$R - CHO + 2e + 2H^+ + \rightarrow RCH_2OH \qquad (3)$$

Reduction of inorganic metallic ions may be represented in a similar way:

$$Cd^{++} + 2e \rightarrow Cd^{\circ}$$
$$Mn^{++} + 2e \rightarrow Mn^{\circ}$$
$$Co^{+++} + 1e \rightarrow Co^{++}$$
$$CO^{++} + 2e \rightarrow Co^{\circ}$$

The question of feed additive metabolites is becoming increasingly important. It is well known (at least with insecticides) that many metabolites are considerably more toxic or harmful than the parent compound. Since the half-wave potential is a characteristic of the compound, the metabolites would likely have a different half-wave potential even though the same group is being reduced. On this basis, it may even be possible to detect and quantitate a residue of the feed additive and its metabolite(s) in the same solution.

Frequently the most important part of microanalytical methods of formulation and residue analysis is the cleanup procedure. Interfering materials must be eliminated before a quantitative measure of the compound can be attained. After the compound has been isolated, the detection system should serve as a qualitative check at the time of quantitation. Ultraviolet and colorimetric methods do this, although both have limitations and require recording spectrophotometers. Trace impurities can lead to inaccurate and misleading information or no results at all. Colorimetric methods, such as those used in feed additive analysis, often are very similar with only a small wavelength shift that may not be apparent unless the complete spectra is recorded and the wavelength absorption maxima observed.

## V. Polarographic applications

The following part of the review deals with compounds now in use as feed additive medicaments for which there are polarographic potentialities (see Table III). References to literature indicate known uses for polarography as methods for feed additives, or as information that seems relevent in the development of a method for medicaments. Commentary is also given on various compounds to indicate where polarographic techniques might be applied.

2-Acetyl-amino-5-nitrothiazole (Acetyl Enheptin) is reducible at the dropping mercury electrode. The polarogram is well defined and is quite sensitive as expected for the reduction of a nitro group. The half-wave potential is just under $-0.5$ volt vs. the saturated calomel electrode which is in the range of other nitro-group-containing compounds. The polarogram can be recorded when the compound is in an aqueous alkaline solution such as sodium hydroxide or ammonium hydroxide.

Acetyl Enheptin may also be chromatographed by gas liquid chromatography (BECKMAN and ALLEN 1962) on a two-foot, $2^{1}/_{2}$ percent SE-30 silicone rubber gum on Analabs ABS support. The compound can be collected and the polarogram recorded. This may be used as a means of cleanup for the sample.

Acetyl-*p*-nitrophenylsulfonamide (APNPS) has been determined polarographically by DAFTSIOS and SCHALL (1962 a and b). Their method utilized A. C. polarography although conventional D. C. equipment can be used.

Table III. *Polarographically active feed additives*

| Common name | Chemical name |
|---|---|
| APNPS . . . . . . . . . . | acetyl-*p*-nitrophenyl sulfanilamide |
| Aureomycin . . . . . . . | chlortetracycline |
| Diethyl stilbestrol . . . . . . | 3,4-bis(*p*-hydroxyphenyl)-3-hexene (nitro derivative) |
| DNBA . . . . . . . . . . | 3,5-dinitrobenzamide |
| Enheptin . . . . . . . . . | 2-amino-5-nitrothiazole |
| Enheptin A . . . . . . . . | 2-acetylamino-5-nitrothiazole |
| Furazolidone . . . . . . . | 3-(5-nitrofurfurylideneamino)-2-oxazolidone |
| 3N4 . . . . . . . . . | 3-nitro-4-hydroxyphenylarsonic acid |
| 4NA . . . . . . . . . . | 4-nitrophenylarsonic acid |
| Nicarbazin . . . . . . . . | 4,4'-dinitrocarbanilide and 2-hydroxy-4,6-dimethyl pyrimidine |
| Nithiazide . . . . . . . . | 1-ethyl-3-(5-nitro-2-thiazolyl)urea |
| Nitrofurazone . . . . . . . | 5-nitro-2-furaldehyde semicarbazone |
| Nitrophenide . . . . . . . | bis(*m*-nitrophenyl)-disulfide |
| Penicillin . . . . . . . . . | benzylpenicillic acid |
| Streptomycin . . . . . . . | — |
| Sulfaquinoxaline . . . . . . | $N'$-(2-quinoxalinyl)-sulfanilamide |
| Terramycin . . . . . . . . | hydroxytetracycline |
| Tran Q. . . . . . . . . | hydroxyzine dihydrochloride |
| Zoalene . . . . . . . . . | 3,5-dinitro-*o*-toluamide |

The half-wave potential reported was $-0.81$ volt vs. the mercury pool. The method was written for the simultaneous determination of acetyl-*p*-nitrophenylsulfonamide and 3,5-dinitrobenzamide, which utilized an aluminum oxide chromatographic column to clean up the sample after acetone extraction. The solvent was aqueous dimethylformamide-tetraethylammonium bromide for both chromatography and polarographic analysis. Recovery of the drugs added to poultry feed was 98 to 99 percent. The standard deviation was reported to be less than two percent. This method should require less time than the separate colorimetric analysis for the two drugs and should be applicable to other feed additives using A. C. or D. C. polarography.

2-Amino-5-nitrothiazole is commonly known as Enheptin. Since Acetyl Enheptin is polarographically active, and a nitro group is present, it should also respond. Existing isolation or cleanup methods could be applied to feed samples containing this compound for analysis by polarography.

Amprolium is the common name for 1-(4-amino-2-*n*-propyl-5-pyrimidinylmethyl)-2-picolinium chloride hydrochloride. Some substituted pyrimidine compounds are reducible, while others are not. No polarographic reference has been found on this compound.

Recently it has been reported that some organic arsenicals are reducible at the DME (DAFTSIOS and SCHALL 1962) which indicated excellent preliminary results with arsanilic acid analysis by polarography. It would be of interest to study carbasone and other organic arsenicals polarographically.

Conversion of the amino group in arsanilic acid to a nitroso group should be possible so that the derivative could be determined polarographically.

A wet digestion or ashing of samples containing cadmium anthranilate leading to inorganic cadmium ion would permit polarographic analysis of this compound. A further cleanup using ion exchange chromatography or dithizone separation followed by polarography has been reported.

Cadmium oxide in feeds could be determined easily by polarography in an acid or ammonical solution, producing well defined waves (KOLTHOFF and LINGANE 1941).

Chlorotetracycline has been studied polarographically (BREZINA and ZUMAN 1958). It is reported that buffers of pH 5.8 to 8.2 were best suited for the determination of chlorotetracycline. A phosphate buffer of pH 5.1 was suitable for the characterization of chlorotetracycline.

GAJAN (1963) reported that for quantitative estimation, diethylstilbestrol could be nitrated, and then polarographed. This compound has been a problem for the feed additive analyst for some time and this method should prove useful as an alternate method for confirmation of its presence and concentration.

The compound 3,5-dinitrobenzamide has been analyzed polarographically by DAFTSIOS and SCHALL (1962b) using A. C. polarography. Conventional D. C. equipment could also be used. They reported on the feasibility of a simultaneous determination of 3,5-dinitrobenzamide and acetyl-p-nitrophenylsulfonamide, and they also discussed the simultaneous determination of DNBA and Zoalene. It should be possible to use conventional chromatographic methods for separation of DNBA from feed extracts followed by polarography for detection and quantitation. Gas chromatography has been used as a means of sample cleanup for this additive (BECKMAN and ALLEN 1962). The material was collected from the gas chromatography using temperature programming (isothermal equipment will also function) and the amount measured by polarography. The column used was a two-foot, $2^1/_2$ percent SE-30 on Analabs ABS 60/70 mesh. It is advisable to turn off the thermal conductivity detector when making a collection to prevent losses of the compound from the hot filaments. Losses of 40 percent of the compound were noted when collections were made with the detector operating.

Degradation of di-n-butyltin-dilaurate to inorganic tin could lead to the indirect analysis of this compound by polarography.

Furazolidone and nitrofurazone in feeds has been analyzed polarographically by MOORE and GUERTAL (1960). MISS and CHIALE (1955) studied the polarograms of nitrofurans in aqueous solutions at pH 9 to 3.2. Polarographic analysis of nitrofuran has also been reported by TATE (1948). He studied nitrofurans in various solvents containing buffer solution. SASAKI (1954) reported that the wave height and diffusion currents for nitrofurazone are not influenced by pH and are constant at pH 4 to 12.

The analytical scheme for furazolidone and nitrofurazone used by MOORE and GUERTAL (1960) included an acetone extraction, evaporation of the acetone, addition of 10 ml. each of DMF and of ammonium citrate and filtering the solution through an aluminium oxide chromatographic column directly into an electrolysis vessel. After removal of oxygen, the polarogram was recorded. Many compounds containing nitro groups, including furazolidone and nitrofurazone, can be polarographed in a DMF ten percent aqueous tetraethylammoniumbromide solvent system.

A furazolidone-nitrofurazone mixture is marketed as Bifuran. The two compounds need to be separated for regulatory purposes. This can be achieved by column chromatography (BECKMAN 1958). Separation of the compounds is made on an aluminum oxide chromatographic column. The mixture is added to the column in a DMF solution, the furazolidone is eluted with the solvent, and the nitrofurazone is later eluted with DMF-alcohol. Following the addition of an electrolyte the two fractions can then be determined polarographically.

Polarography has been applied to the analysis of Tran Q (hydroxyzine dihydrochloride) (PFIZER 1958). The procedure uses a treated carbon electrode (see GAYLOR et al. 1957 for further discussion of this electrode). The polarogram is recorded from $+1.35$ volts with a span of 1.5 volts. The diffusion current is measured at its peak which is reported to occur at about $+0.75$ volt. Sample cleanup is effected after an isopropyl alcohol extraction by partitioning the Tran Q between aqueous acid and chloroform, and after adjusting the pH of the aqueous phase to 10.5 to 11.0, re-extraction by chloroform. The final cleanup is to extract the hydroxyzine from the chloroform with acid solution. The solution is then buffered, deoxygenated, and the polarogram recorded.

Nicarbazin is a molecular complex of 4,4'-dinitrocarbanilide and 2-hydroxy-4,6-dimethylpyrimidine. 2,4,6-Trisubstituted pyrimidines are not reducible (BREZINA and ZUMAN 1958, CAVALIERI and LOWY 1952). The dinitro part of the complex is reducible and can be determined polarographically (DAFTSIOS and SCHALL 1962). By using the existing cleanup method followed by polarographic detection, an alternate method of analysis would be available for confirmation of results.

Nicotine is reported to give a catalytic wave in the pH region of 5 to 10 with a maximum at pH 6. For quantitative analysis, pH 8 is useful for $10^{-4} M$ concentrations. The wave is reported to be ill-defined in acid solutions and masked by the much larger catalytic wave. In alkaline solutions, the catalytic wave is small and the reduction wave almost coincides with the wave for the supporting electrolyte.

Nithiazide is a condensation of ethyl isocyanate and 2-amino-5-nitro-thiazole. Since 2-amino-5-nitrothiazole (Enheptin) can be polarographed, the derivative, nithiazide should be reducible. The half-wave potentials of Enheptin, Acetyl Enheptin, and nithiazide are probably similar, although there is generally enough difference caused by different substitution to make a positive identification. This is especially true for A. C. polarography.

Bis(m-nitrophenyl)-disulfide is known as nitrophenide. DAFTSIOS and SCHALL (1962) reported on the feasibility of A. C. polarographic determination of the compound. As in other feed materials, some cleanup would be necessary, but it should be possible to utilize existing extractions and isolation procedures.

Another compound used extensively in poultry feeds is 3-nitro-4-hydroxyphenylarsonic acid and it has been studied polarographically by several workers (DAFTSIOS and SCHALL 1962 a, WAWZONEK 1958). The usual method of determination in feed samples is by the Gutzeit arsenic analysis. A specific procedure utilizing polarography is possible since the nitro group is reducible. Further work is needed on the separation of this compound from the other nitro-group-containing feed additives. 4-Nitrophenylarsonic acid has been studied polarographically. The wave is similar to other mono-nitro groups containing aromatic compounds. Polarography could be used as a specific analytical method for this compound.

Hydroxytetracycline (terramycin) is reduced in two two-electron steps which most likely corresponds to the reduction of the two-unsaturated carbonyls. It is reported that chlorotetracycline and hydroxytetracycline differ in their behavior in a solution containing boric acid. For the determination of hydroxytetracycline, a pH of 5 to 6.5 is best. For simultaneous determination of hydroxytetracycline and chlorotetracycline, a phosphate buffer of pH 8.1 is most suitable (BREZINA and ZUMAN 1958).

Conversion of the amino group in $p$-aminobenzoic acid to a nitroso group should lead to a polarographically active compound. The oxidation of the amino group to a nitroso group is possible with a $K_2S_2O_8 - H_2SO_4$ mixture (HORNING 1955).

Penicillin can be determined by several polarographic methods. One depends on the bacteriostatic suppression of the respiration of staphylococcus by penicillin. The oxygen need is measured by recording the oxygen wave after 90 to 120 minutes. The higher the concentration of penicillin, the smaller the need for oxygen and the oxygen wave will be higher (BREZINA and ZUMAN 1958).

Another indirect method utilizes hydrolysis in which substances are formed that contain sulfhydryl groups that are capable of catalyzing the reduction of hydrogen ions in Brdička's cobalt solution. This method is sensitive but not specific.

Penicillin can also be determined by forming a stable nitroso compound. Procain penicillin can be analyzed for both components.

Penicillin can be hydrolyzed with dilute alkali to penicilloic acid and converted with dilute acid to penicillamine which contains a free sulfhydryl group that is polarographically active and gives a catalytic wave in Brdička's solution. By this method, 0.5 international unit in 1.0 ml. can be determined with an accuracy similar to that of the biological test (BREZINA and ZUMAN 1958).

The polarographic behavior of streptomycin closely resembles that of aliphatic aldehydes (BREZINA and ZUMAN 1958). Good agreement between microbiological and polarographic analysis is reported (LEVY et al. 1946). The height of the wave for streptomycin changed with pH in a manner similar to that observed for formaldehyde and reached a maximum at pH 13 (BRICKER and VAIL 1951). The most suitable supporting electrolyte is reported to be tetramethylammonium hydroxide at pH 13.6 to 13.8.

Sulfaquinoxaline (2-sulfanilamidoquinoxaline) can be determined polarographically giving a well-defined wave in buffered alkaline supporting electrolytes. Use of existing cleanup procedures should lead to a sensitive alternate method of analysis.

3,5-Dinitro-o-toluamide (Zoalene) is polarographically similar to 3,5-dinitrobenzamide; however, the two can be distinguished by their half-wave potentials. DAFTSIOS and SCHALL (1962 a) suggested the use of dimethyl-formamide-ten percent aqueous tetraethylammonium bromide supporting electrolyte. It should be noted that conventional polarography can be substituted for A. C. polarography, although the latter is probably better for mixtures where additional resolution is needed. Zoalene can be chromatographed by GLC under conditions similar to those for 3,5-dinitrobenzamide (BECKMAN and ALLEN 1962), and this technique can be used for sample cleanup and separation, followed by polarography. Recovery of the compound is possible since the ultraviolet and infrared spectra before and after gas chromatography are identical (see Figs. 1 and 2). It is advantageous to collect the compound from the gas chromatograph with the detector off to avoid loss from the heat of the filaments. Other types of cleanup such as adsorption, partition, paper- or thin-layer chromatography may be employed for isolation of 3,5-dinitro-o-toluamide. The technique of using thin-layer chromatography for infrared determination on organic compounds has been reported (NASH et al. 1963). With modification this procedure should be applicable to the polarography of compounds such as Zoalene.

## Summary

A brief historical summary of the beginning of the use of chemicals in feeds is given followed by a brief comment on the present buildup in the use of medicaments. A need for new and alternate methods of analysis is described, and it is pointed out that regulatory laboratories responsible for accurate analysis of feed additives are not taking full advantage of all methods and techniques available. A review of methods available with some commentary on their usefulness is given as a background to the newer methods of gas chromatography, infrared spectroscopy and polarography.

Based on the literature and preliminary investigations undertaken by the authors, polarography can be adapted as a supplementary and most important means of analysis of medicaments in animal and poultry feeds. It can be used as a qualitative and quantitative procedure and, since the major portion af the sample is not destroyed during the analysis, it can be recovered, if necessary, for other uses. In common with all analytical procedures for feed additive and/or residue analysis, efficient cleanup of sample is necessary. A combination of GLC cleanup plus polarographic analysis can be an efficient means of feed additive analysis.

## Résumé *

Un aperçu historique de l'origine de l'usage des produits chimiques dans les aliments pour animaux est présenté et suivi d'un bref commentaire sur

---

* Traduit par S. DORMAL VAN DEN BRUEL.

l'état actuel de l'emploi des médicaments. Le besoin de méthodes d'analyse nouvelles et de remplacement est exposé, et il ressort que les laboratoires officiels qui ont la responsabilité de l'analyse précise des additifs alimentaires ne tirent pas pleinement parti de toutes les méthodes et techniques disponibles. Une revue des méthodes existantes, accompagnée de commentaires sur leur utilité, sert de base à l'examen des méthodes d'analyse plus récentes par chromatographie gazeuse, spectroscopie infra-rouge et polarographie.

D'après la littérature et les recherches préliminaires entreprises par les auteurs, la polarographie peut être adaptée comme moyen supplémentaire très important d'analyse des médicaments dans les aliments pour animaux et volaille. Elle peut être utilisée comme procédé qualitatif et quantitatif et, comme la majeure partie de l'échantillon n'est pas détruite au cours de l'analyse, celui-ci peut être récupéré, si nécessaire, pour d'autres usages. De même que pour les autres procédés d'analyse des additifs alimentaires et (ou) pour l'analyse des résidus, une purification poussée de l'échantillon est nécessaire. Une combinaison du procédé de purification par chromatographie gaz-liquide et du dosage polarographique peut être un moyen efficace d'analyse des denrées alimentaires.

## Zusammenfassung *

Es wird zunächst eine kurze historische Zusammenfassung über die erste Anwendung von Chemikalien in Futtermitteln gegeben, darauf folgt ein kurzer Kommentar über den gegenwärtigen Stand des Einsatzes von Medikamenten. Die Notwendigkeit für neue, alternative Analysenmethoden wird beschrieben und es wird darauf hingewiesen, daß Kontroll-Laboratorien, die für exakte Analysen von Futterzusatzstoffen verantwortlich sind, die Vorteile aller verfügbaren Methoden und Techniken nicht voll nutzen. Als Hintergrund für die neueren Methoden der Gaschromatographie, Infrarotspektroskopie und Polarographie wird eine Übersicht über die verfügbaren Methoden mit einigen Kommentaren hinsichtlich ihrer Verwendungsmöglichkeiten gegeben.

Wie aus der Literatur und vorläufigen Untersuchungen der Autoren hervorgeht, läßt sich die Polarographie zu einem zusätzlichen Analysenverfahren von größter Bedeutung für die Bestimmung von Medikamenten in Vieh- und Geflügelfutter ausbauen. Sie kann als qualitatives und quantitatives Verfahren eingesetzt werden, und da der überwiegende Teil der Probe bei der Analyse nicht zerstört wird, kann er zurückgewonnen und gegebenenfalls für andere Zwecke weiterverwandt werden. Wie bei allen analytischen Verfahren für Zusatzstoffe und/oder Rückstände, ist auch hier eine ausreichende Vorreinigung der Analysenprobe notwendig. Eine Kombination aus gaschromatographischer Reinigung und polarographischer Analyse kann ein wirkungsvolles Verfahren zur Analyse von Futtermittelzusätzen darstellen.

---

* Übersetzt von H. FREHSE.

## References

BAERSTEIN, H. D.: Photometric determination of benzene, toluene, and their nitro derivatives. Ind. Eng. Chem., Anal. ed. 15, 251 (1943).

BARNARD, R. D.: Simple color reaction for piperazine. J. Amer. Pharm. Soc., Sci. ed. 36, 224 (1947).

BECKMAN, H. F.: Microdetermination of the medicaments furazolidone and nitrofurazone. J. Agr. Food Chem. 6, 130 (1958).

— Excerpts from dissertation, "Colorimetric microanalysis of several organic feed additives". Agr. and Mech. College of Texas. January 1959 a.

—, and J. DE MOTTIER: Colorimetric determination of 3,5-dinitrobenzamide in feeds by the Janovsky reaction. J. Agr. Food Chem. 7, 280 (1959).

—, and L. S. FELDMAN: Colorimetric determination of cadmium anthranilate in feedstuffs. J. Agr. Food Chem. 7, 350 (1959).

— — Feed additive analysis: microanalysis of piperazine. J. Agr. Food Chem. 8, 227 (1960).

—, and P. T. ALLEN: New development in drug additive analysis. Proc. 10th Ann. Meeting, Amer. Assoc. Feed Microscopists. P. 129. Chicago, June 1962.

VON BITTO', B.: Reaction of aldehydes and ketones with metadinitroaromatics to produce colored products. Ann. 269, 377 (1892).

BOST, R. W., and F. NICHOLSON: A color test for the identification of mono-, di-, and trinitro compounds. Ind. Eng. Chem., Anal. ed. 7, 190 (1935).

BREZINA, M., and P. ZUMAN: Polarography in medicine, biochemistry, and pharmacy. English ed. New York: Interscience 1958.

BRICKER, C. E., and W. A. VAIL: A polarographic investigation of the alkaline decomposition of streptomycin. J. Amer. Chem. Soc. 73, 585 (1951).

BRÜGGEMANN, J., K. BRONSCH, H. HEIGNER, and H. KNAPSTEIN: Feed additives: nitrofurazone determination in poultry feed by phenylhydrazine or alkali reaction. J. Agr. Food Chem. 10, 108 (1962).

BUZARD, J. A., V. R. ELLS, and M. F. PAUL: Determination of nitrofurazone and furazolidone in feeds. J. Assoc. Official Agr. Chemists 39, 512 (1956).

CAMPBELL, K. N., B. K. CAMPBELL, and S. J. PATELSKI: p-Metoxyphenylisocyanate as a reagent for the identification of amines. Proc. Indiana Acad. Sci. 53, 119 (1943); Chem. Abstr. 39, 881 (1945).

CASTIGLIONI, A.: Determination of piperazine. Z. Anal. Chem. 117, 25 (1939); Ibid. 119, 118 (1940); Chem. Abstr. 33, 6756 (1939); Ibid. 34, 4699 (1940).

—, and M. NIVOLI: Paper chromatographic separation of piperazine with hexamethylenetetramine. Z. Anal. Chem. 138, 187 (1953); Chem. Abstr. 47, 5848 (1953).

CAVALIERI, L. F., and B. A. LOWY: A polarographic study of pyrimidines. Arch. Biochem. Biophys. 35, 83 (1952).

CAVETT, J. W.: Colorimetric method for determination of 3-nitro-4-hydroxyphenylarsonic acid in feed. J. Assoc. Official Agr. Chemists 39, 857 (1956 a).

— Assay of acetyl-(p-nitrophenyl)sulfanilamide in feeds. J. Assoc. Official Agr. Chemists 39, 964 (1956 b).

— Colorimetric method for the determination of 4-nitrophenylarsonic acid in feed. J. Assoc. Official Agr. Chemists 39, 967 (1956 c).

— Assay for dinitrodiphenylsulfonylethylenediamine in feed. J. Assoc. Official Agr. Chemists 39, 969 (1956 d).

—, and J. P. HEOTIS: Report on piperazine in feeds. J. Assoc. Official Agr. Chemists 41, 323 (1958).

— — Report on 3,5-dinitrobenzamide in feeds. J. Assoc. Official Agr. Chemists 42, 239 (1959).

CRIPPA, A.: Chloranil as a reagent for the visualization of various substances in paper chromatography. Ist. botan. univ. Lab. crittogram., Pavia, Atti 10, 173 (1953); Chem. Abstr. 48, 3187 (1954).

CROSS, A. H. J., R. A. HENDEY, and S. G. E. STEVENS: The determination of nitrofurazone in feed mixes. Analyst 85, 657 (1960).

DAFTSIOS, A. C., and E. D. SCHALL: Applications of A.C. polarography to the analysis of drugs in feeds. J. Assoc. Official Agr. Chemists 45, 278 (1962).

DAFTSIOS, A. C., and E. D. SCHALL: The simultaneous determination of 3,5-DNBA and APNPS by A. C. polarography. J. Assoc. Official Agr. Chemists 45, 291 (1962).

Dow Chemical Company: A.C.D. information bulletin 106. Midland, Michigan 1956.

ELLS, V. R., E. S. McKAY, and H. E. PAUL: Determination of nitrofurazone in feeds. J. Assoc. Official Agr. Chemists 36, 417 (1953).

ENGLISH, F. L.: Colorimetric determination of certain dinitro aromatics. Anal. Chem. 20, 745 (1948).

FEINSTEIN, L.: A color test for pyrethrins and allethrin. Science 115, 245 (1952).

FISCHER, W.: Macroanalysis of various polynitro aromatic compounds using the Janovsky reaction. Z. Anal. Chem. 131, 192 (1950).

Fisher Scientific Company: Technical data sheet TD-142. 1960.

FOSTER, R., and R. K. MACKIE: Interaction of electron acceptors with bases — V: The Janovsky and Zimmermann reactions. Tetrahedron 18, 1131 (1962).

FOUCRY, M.: Quinone, a reagent for amines. J. pharm. chim. 20, 116 (1934); Chem. Abstr. 28, 7195 (1934).

FUSON, R. C.: Advances in organic chemistry. P. 370. New York: Wiley 1950.

GAJAN, R. J.: Applications of polarography for detection and determination of pesticides and their residues. Amer. Chem. Soc. 144th Ann. Meeting. Los Angeles, California, March 1963.

GAYLOR, V. F., L. CONRAD, and J. H. LANDERL: Use of a waximpregnated graphite electrode in polarography. Anal. Chem. 29, 224 (1957).

Hess and Clark, Inc.: Unpublished methods. Ashland, Ohio 1954.

HOFFMAN, H. H.: Private communication. 1956.

HORNING, E. C.: Organic syntheses. Vol. 3, P. 334. New York: Wiley 1955.

HORWITZ, W., ed.: Official methods of analysis of the association of official agricultural chemists. Ninth ed. Washington, D. C. 1960.

JANOVSKY, J. V., and L. ERB: Color reactions of nitro- and azobenzenes in acetone. Ber. 19, 2155 (1886).

— Color reactions of dinitro derivatives of phenols, halobenzenes, and alkyl-benzenes. Ber. 24, 971 (1891).

Jefferson Chemical Company: "Piperazine" tech. bull. Houston, Texas 1956.

KLEIN, A. K., and H. WICHMANN: Determination of cadmium with dithizone in foods. J. Assoc. Official Agr. Chemists 28, 257 (1945).

KOLTHOFF, I. M., and J. J. LINGANE: Polarography. P. 269. New York: Interscience 1941.

— — Polarography. Vol. 2, chapter XXIX. New York: Interscience 1952.

LENG, M. L.: Chemical determination of piperazine as the dihydrochloride in feeds and concentrates. J. Assoc. Official Agr. Chemists 40, 1059 (1957).

LEVY, G. B., P. SCHWED, and J. W. SACKET: Polarographic analysis of streptomycin. J. Amer. Chem. Soc. 68, 528 (1946).

Limecrest Research Laboratory: Unpublished report. New Jersey 1956.

MADER, J. W.: A source of interfering color in the colorimetric analysis of arsanilic acid. J. Assoc. Official Agr. Chemists 39, 321 (1956).

MERWIN, R. T.: Symposium on medicated feeds. Ed. H. WELCH and MARTIN-IBANEZ. New York: Medical Encyclopedia 1956.

MILLER, D. M.: A method of recording A. C. polarograms on a conventional D. C. polarograph. Can. J. Chem. 34, 942 (1956); also, see Can. J. Chem. 35, 1589 (1957).

MISS, A., and R. CHIALE: Polarographic determination of 5-nitro-2-furaldehyde semicarbazone. Rev. Chim. (Bucharest) 6, 41 (1955); Chem. Abstr. 49, 8744 (1955).

MOORE, H. P., and C. R. GUERTAL: Polarographic determination of nitrofurans in feeds. J. Assoc. Official Agr. Chemists 43, 308 (1960).

NAGASAWA, K., and S. OHKUMA: J. Pharmacol. Soc. Japan 74, 410 (1954).

NASH, N., P. ALLEN, A. BEVENUE, and H. BECKMAN: A technique for the recovery of compounds from thin layer chromatograph strips for infrared analysis. J. Chromatog., in press (1963).

Norwich Pharmacal Company: Unpublished data. Norwich, New York 1950.

PAGE, J. E.: Chemical and physical methods for penicillin assay. Analyst 73, 197 (1948).

*Pfizer, Charles and Company:* Determination of Tran Q in feeds. Brooklyn, New York 1958.

REITZENSTEIN, J., and G. STAMM: Mechanism of the base-catalyzed reaction of acetone with metadinitrochlorobenzene. J. Prakt. Chem. 81, 168 (1910).

RUDOLPH, O.: Base-catalyzed color products of various di- and trinitro compounds with acetone. Z. Anal. Chem. 60, 239 (1921).

SALTZMAN, B. E.: Colorimetric microdetermination of cadmium with dithizone. Anal. Chem. 25, 493 (1953).

SAZAKI, T.: Polarographic study of nitrofuran derivatives I. Polarographic behavior of 5-nitrofurfuraldehyde semicarbasone; II. Reduction potential of nitrofuran derivatives and nitrobenzene analogs. Pharm. Bull. (Japan) 2, 99 (1954); Chem. Abstr. 49, 11460 (1955).

SCOTT, W. W.: Scott's standard methods of chemical analysis. Fifth ed., vol. 1, p. 202 and vol. 2, p. 1516. New York: Van Nostrand 1939.

STONE, L. R.: Unpublished data.

TATE, I.: Polarography of nitrofuran derivatives. J. Pharm. and Chem. (Japan) 20, 38 (1948); Chem. Abstr. 45, 3257 (1951).

VITTE, G., and M. COUSTOU: Determination of small amounts of piperazine and mixtures. Bull. trav. soc. pharm. Bordeaux 81, 100 (1943); Chem. Zentr. II, 457 (1944); Chem. Abstr. 40, 7082 (1946).

WAWZONEK, S.: Organic polarography. Anal. Chem. 30, 661 (1958).

# The potential of fluorescence for pesticide residue analysis

By

D. MacDougall *

With 1 figure

## Contents

## I. Introduction

As a technique for the determination of pesticide residues, fluorescence is not as generally applicable as some modern methods such as vapor phase chromatography. However, it does have certain advantages over the classical methods of analysis in cases where it can be used. The possible uses of this technique for residue purposes have not been thoroughly explored. It is important that the residue chemist consider all possible techniques available and use whichever is most readily applicable to a given problem. Therefore, in this paper the author will attempt to explore the areas of residue analysis in which fluorescence should be applicable. No attempt is made to make an exhaustive review of fluorescence methods in general.

The use of fluorescence methods for the determination of pesticide residues has been discussed in the first volume of this series (MacDougall 1962). In this paper, only a brief reference will be made to material covered in the previous review.

---

* Chemagro Corporation, Kansas City, Missouri.

## II. Advantages and disadvantages

The principal advantage of fluorescence methods over classical methods such as colorimetry is sensitivity. As was pointed out in the previous paper (MacDougall 1962), fluorescence methods may be up to one — thousand times as sensitive as colorimetric procedures. The increased sensitivity allows the residue chemist great flexibility in setting up his procedure. When the full sensitivity potential of the method is not required, proportionately smaller samples may be used with a consequent saving in amounts of reagents. This also allows a great saving in time, especially where solvent evaporations are involved.

One disadvantage of the photofluorometric method is that it is limited in its applicability. In spite of this, the actual potential of this technique for pesticide residue analysis has not been nearly realized. It is the principal purpose of this paper to discuss the possible scope of fluorescence in pesticide residue analysis.

The second disadvantage in using the fluorescence method is that many naturally occurring materials fluoresce and, therefore, considerable difficulty may be experienced in cleanup procedures in order to obtain background values which are sufficiently low to allow utilization of the sensitivity of which the method is capable. There is no easy way of getting around this problem. The actual conditions for cleanup for any particular compound and its metabolites in a given substrate have to be carefully worked out. However, the problems involved in this respect have been overcome in a number of cases and are by no means insurmountable.

Another property of fluorescent materials which may be considered an advantage as well as a disadvantage is the fact that the fluorescence intensity is dependent upon a variety of environmental conditions. The effects of solvent and pH are particularly important in this connection. The effect of solvent on the intensity of fluorescence has already been reported (MacDougall 1962). It was shown that the relative fluorescence of anthranilic acid varies almost two-hundred-fold depending on which solvent is used. It may well be that these results were due in part to the fact that reagent grade solvents which had not been specially purified were used. The presence of a small amount of water or of a polar solvent in a non-polar solvent may greatly affect the intensity of fluorescence.

pH also has a very marked effect on fluorescence. This was illustrated in the previous paper (MacDougall 1962).

Concentration quenching is a serious problem. This results in the relationship between fluorescence and concentration being linear only over a limited concentration range. Great care must be exercised to insure that measurements are made in the range in which linearity exists.

These difficulties in cortrolling the external and environmental factors often mean that an internal standard technique must be used with all samples. This consists of adding increment amounts of the compound to the sample to be analyzed. Measurements are then made and it is assumed that the difference in reading between the sample alone, and the sample to which the standard has been added, is due to the addition of the standard.

By extrapolation, it is possible to calculate the concentration of material in the unknown.

Because of the extreme effects of solvent type, solvent purity, pH, etc. on fluorescence intensity, it is necessary to do a very careful study of any compound before assuming that it does not fluoresce. This should include examination in a number of different solvent systems at different pH's, etc.

Some of the residue determinations for which the fluorescence method has been used were listed in the previous paper (MacDougall 1962).

## III. Possible scope

Now let us turn to a consideration of the possible scope of photo-fluorometry in the analysis of pesticide residues. Many of the best possibilities have never been investigated at all.

The general approaches that may be used are:

1. Direct measurement of fluorescence.
2. Conversion to a highly fluorescent compound.
   a) By direct conversion to a fluorescent material.
   b) By coupling the parent compound to form a fluorescent material.
   c) By coupling a portion of the molecule to form a fluorescent derivative.
3. Indirect methods in which a chemical property of the compound is utilized to form a fluorescent material.
4. Indirect methods in which a chemical property of a compound can be used to form a fluorescent chelate.

Each of these approaches will be considered separately.

### a) Direct measurement of fluorescence

Obviously, this method applies only to materials which are fluorescent to begin with. Some examples of the types of materials which are fluorescent are listed below. There are undoubtedly many others whose fluorescence needs only to be investigated.

> Polynuclear compounds
> Steroids
> Gibberellic acid
> Diethylstilbesterol
> Warfarin
> Indoleacetic acid
> Naphthalene acetic acid

The general fluorescence of polynuclear materials has been reported in many instances (Druckrey et al. 1959). It should be possible to use this property to measure the fluorescence of residues resulting from application of oils to plants as the oils used should undoubtedly contain polynuclear materials.

A tremendous amount of work has been done on the fluorescence of steroids. The *Silber* reaction which is based on the fluorescence of steroids in sulfuric acid has been studied by a number of workers (Hosoi 1958, Inai 1960). This technique has been applied to a number of drugs, e. g. cortisone, hydrocortisone, etc. (Braunsberg and James 1960).

The fluorescence of gibberellic acid in sulfuric acid has been described by Arison et al. (1957) and by Theriault et al. (1961).

The fluorescence of coumarin derivatives is well known. Ichimura (1959) described a procedure based on the fluorescence of warfarin in ethanol at pH 10.

Hornstein (1958) mentioned the fluorescence of indoleacetic acid and naphthalene acetic acid some time ago.

Residue methods for diethylstilbesterol in beef liver and in cattle feed have been described (Cheng and Burroughs 1955, Munsey 1958, Goodyear and Jenkinson 1961).

### b) Direct conversion to fluorescent compound

Both Co-Ral and Guthion are converted to highly fluorescent materials by alkali and heat (Anderson et al. 1959, Adams and MacDougall 1961). The reactions involved in these hydrolyses were shown in the previous paper (MacDougall 1962).

Another interesting reaction which falls in this category is that of phthalic acid on heating with resorcinol and sulfuric acid to form fluorescein (Chomse and Arend 1959, Thommes and Leininger 1960). Several pesticides could be determined by this method. Examples are dimethyl phthalate and Alanap-1. In this determination, a strongly alkaline solution is used because fluorescein fluorescence is not pH-dependent above pH 6. Resorcinol and sulfuric acid give a deep blue fluorescent complex which is pH-dependent and does not fluoresce in alkaline solution.

### c) Coupling parent compound to form a fluorescent derivative

**Amines.** In some cases the parent compound may be coupled to form a fluorescent material. Condensation of agmatine (Cohn and Shore 1961) and histamine (Shore et al. 1959) with o-phthalaldehyde in alkaline solution have been described. However, the reactions are different as acidification of the histamine fluorphor stabilizes and intensifies the fluorescence while in the case of agmatine acidification leads to complete loss of fluorescence. This may indicate that in the latter case the reaction does not proceed beyond a Schiff-base type compound, while the histamine reaction product undergoes further change. A suggested reaction with histamine (Shore et al. 1959) is shown below. Subsequent rearrangement and air oxidation may follow the reaction shown.

o-Phthalaldehyde     Histamine                              Fluorescent

There are several other reactions for alipathic amines which are of interest. One of these (Baker et al. 1952) has recently been investigated by Dr. Adams' group in our laboratory. In this procedure, o-acetoacetylphenol is used for the determination of primary aliphatic amines. This reaction was studied in our laboratory as a possible means of determination of residues of N-methyl carbamates which can be hydrolyzed to liberate methyl amine. It has been reported (BAKER *et al.* 1952) that this reaction reaches 24 percent yield in two days and the product has a strong, greenish-yellow fluorescence in the ultraviolet. The reaction is shown below.

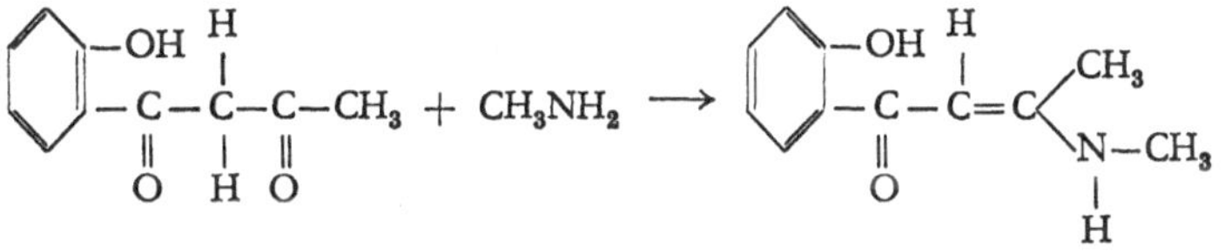

o-Acetoacetylphenol                                    Fluorescent

At residue levels, no reaction was observed with secondary amines, primary aromatic amines, ammonia or amino acids. Thus, this lends a considerable amount of specificity to the reaction. The reaction does not take place in the presence of water and therefore it was found to be necessary to add a water scavenger in order to get the reaction to occur. For this purpose, 2,2-dimethoxy-propane was used. The usual procedure for determination of this type would involve the hydrolysis of the carbamate in alkali, followed by distillation of the methyl amine into a trap containing a dilute solution of hydrochloric acid in ethanol. The contents of the trap would then be evaporated to a small volume and to this would be added a milliliter of a solution of the reagent in dimethoxypropane together with a small amount of pyridine. The purpose of adding the pyridine is to

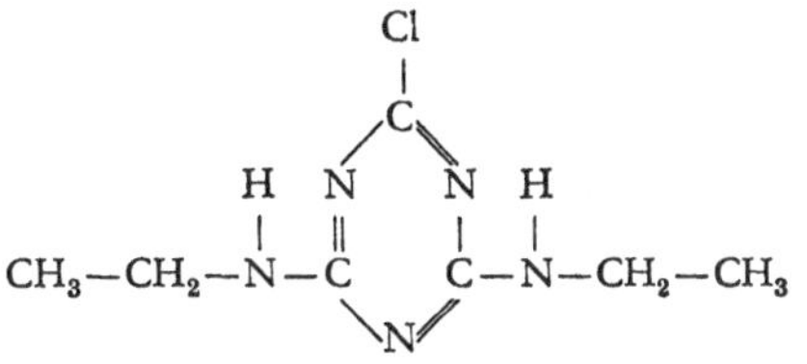

Simazine 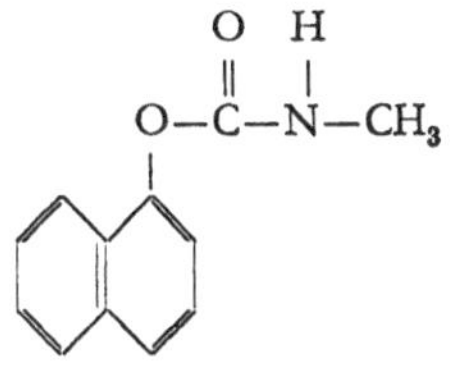                                    Sevin

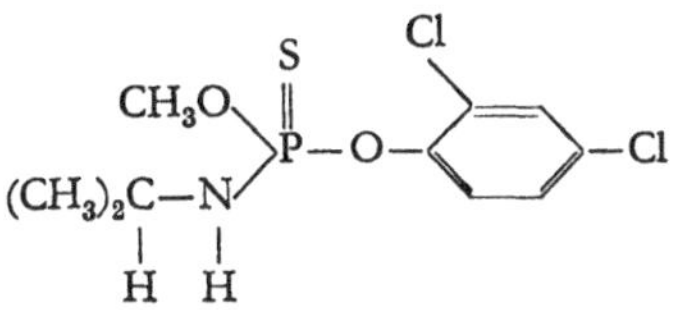

Zytron 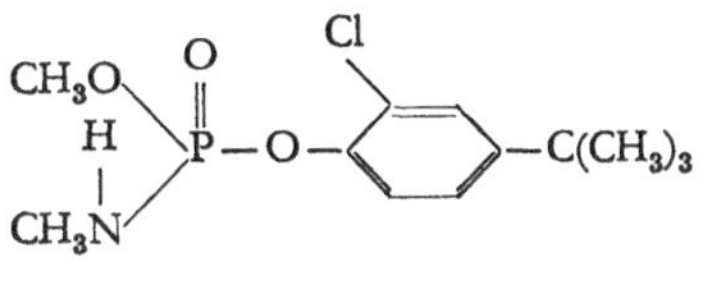                                    Ruelene

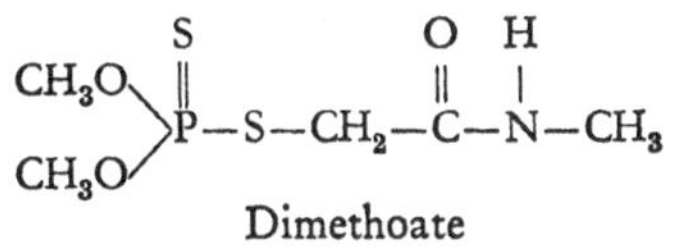

Dimethoate

neutralize the hydrochloric acid present. The samples would then be heated to 60° C. in a water bath for a period of about 16 hours at which time they would be cooled and aliquots removed for fluorescence measurements.

The fluorescence of the reaction product is proportional to the concentration of methyl amine in the range up to 6 $\mu$g./ml. of solution. Additional clean-up work has to be done before this method can be applied for highly sensitive residue determinations. However, it appears promising at this time.

Several pesticides to which this type of procedure could probably be applied are listed below.

**Catechols.** A great deal of work has been done on fluorometric methods for the determination of catechol amines. One of these methods was described originally by WEIL-MALHERBE and BONE (1952). The reaction consists of oxidation and coupling with ethylene diamine to form a fluorescent derivative. It has been shown that this reaction occurs with a wide variety of catechol derivatives including catechol itself. Thus, any compound containing the catechol nucleus could possibly be determined in this way. The reaction with adrenalin is shown below.

Adrenaline

Fluorescent

**Aliphatic acids.** BARR (1948) has described photofluorometric methods for the determination of succinic and malic acids in apples. The reaction is similar to that already mentioned for phthalic acid (THOMMES and LEININGER 1960). FIEGL (1956, pp. 233—234) has reported that dicarboxylic acids with carboxyl groups in the 1,2- or 1,4-positions or their esters, anhydrides, or imides can be determined by this method. It is obvious that this might be a possible method for the dicarboxylic acid formed on hydrolysis of malathion.

Malic acid can also be determined by reaction with $\beta$-naphthol (SORENSEN and MATZKE 1958).

A more general method for aliphatic acids has been described for detection on paper chromatograms (PESEZ and FERRERO 1957). The chromatogram of the acid is developed with butanol saturated with ammonia. The ammonia is allowed to evaporate and the residue is treated with a solution containing the zinc-8-quinolinol complex. In the presence of the aliphatic ammonium salt, a highly fluorescent spot is formed. This method might easily be modified for solutions. Many pesticides could be made to yield aliphatic acids on hydrolysis.

An interesting procedure for fluorometric determination of acetone has been described by BRANDT and CHERONIS (1961). This is accomplished by reaction of acetone with 2-diphenylacetyl-1,3-indandione-1-hydrazone to form an azine. The reaction with acetone is shown below. The sealed-tube reaction between acetone in chloroform and the hydrazone is quite rapid and proceeds to completion in ten minutes at 100° C.

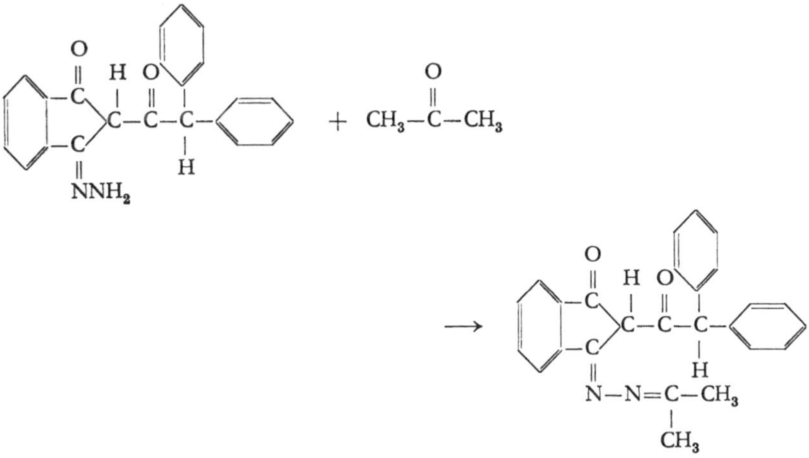

Fluorescent

Similar reactions for formation of fluorescent azines has been described by SAWICKI and STANLEY (1960). Aldehydes such as glyoxal, pyruvaldehyde, or salicylaldehyde give the reaction.

It is obvious that these reactions have wide application in the field of pesticide residue analyses.

FIEGL (1956, pp. 220—222) has described procedures for o-hydroxy-aldehydes and ketones based on formation of aldazines and ketazines.

CONN and DAVIS (1959) have measured the fluorescence characteristics of guanidinium compounds with ninhydrin. This method can be used for determinations with concentrations in the order of 0.07 $\mu$g./ml. This procedure could probably be applied to the determination of dodine residues.

## IV. Indirect methods in which a chemical property of a compound is utilized to form a fluorescent material

An example of this type of procedure is the method described by GEHAUF and GOLDENSON (1957) for the estimation of the nerve gases sarin and tabun. In this procedure, a reagent solution made up of indole and sodium perborate in acetone-water is used. When this is treated with a small amount of a nerve gas, a brilliant green fluorescence appears when examined in ultraviolet light. In this procedure, perborate oxidizes the nerve gas to a peracid which in turn oxidizes indole to indoxyl which is highly fluorescent. It is obvious that this reaction can be applied to organo-phosphates such as Dimefox, Mipafox, and Pestox XIV. It is possible that the procedure could be applied to other organophosphorus compounds as well, if a careful study of reaction conditions were made.

## V. Indirect methods involving formation of a metal chelate

The method suggested by HANKER *et al.* (1958) for determination of mercaptans has been applied in our laboratory (LOEFFLER and MacDougall 1960) to the determination of DEF[1] residues. The basic reactions involved were reported previously (MacDougall 1962). Butyl mercaptan is distilled into a solution of a palladium chelate of 8-hydroxy-5-quinoline sulfonic acid. The mercaptan ties up part of the palladium freeing a corresponding amount of the complexing agent. On addition of magnesium chloride a fluorescent magnesium chelate is formed. The palladium chelate is not fluorescent.

This method can be applied to any other compounds liberating mercaptans or hydrogen sulfide on hydrolysis. The procedure is also applicable to Folex, the phosphite homolog of DEF.

## VI. Analysis of complex systems

In pesticide residue analyses, one may be confronted with the problem of detection of closely related compounds in the presence of the major constituent.

In our own laboratory, the determination of ferron in chlorferron is of interest. The structure of there compounds is shown below.

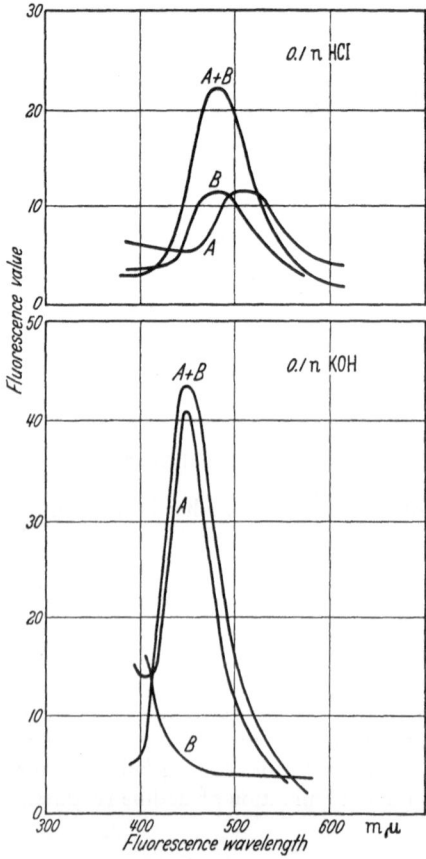

Fig. 1. Effects of acid and base on fluorescent spectrum of ferron (*A*) and chlorferron (*B*); *A*+*B* = mixture at equal concentrations

Chlorferron

Ferron

[1] Trademark, Chemagro Corp.

The curves in Fig. 1 show effect of acid and base on the fluorescence spectrum of ferron and chlorferron. In this figure, $A$ is ferron, $B$ is chlorferron and $A+B$ is a mixture at equal concentrations. Individual components are at the same concentrations as they are in the mixture and all are at 0.5 $\mu$g./ml. It is obvious from these results that the presence of chlorine added to the ferron has a tremendous effect on the fluorescence. This is undoubtedly due to the ease with which the lactone ring is opened.

Another example of the same type is described by THOMMES and LEININGER (1958) for o- and m-hydroxy benzoic acids in mixtures. The procedure is based on the fact that at pH 5.5 only the ortho-isomer is fluorescent and at pH 12 both isomers are fluorescent.

HERCULES and ROGERS (1958) have described a method for the analyses of 1- and 2-naphthols fluorometrically. This is accomplished by measurement of the fluorescence in 0 to 1 $M$ sodium hydroxide at 480 and 426 m$\mu$., respectively.

## VII. Conclusions

In summary, it must be re-emphasized that this discussion is not intended to be a complete review of all the literature on fluorescence or of all the possible applications to pesticides. However, it is hoped that it will serve to indicate the general scope of the technique as applicable to pesticide residues.

## Summary

The uses which have been made of fluorescence for pesticide residue analyses are reviewed, and new areas to which the technique should be applicable are explored. The advantages and limitations of fluorescence techniques, as well as the precautions required in reagent preparation and sample preparation and cleanup, are also reviewed.

## Résumé *

Après une revue d'ensemble des utilisations de la fluorescence antérieurement proposées pour l'analyse des résidus de pesticides sont examinées les possibilités d'application de cette propriété à de nouveaux problèmes. Les avantages et les limites des techniques de fluorescence, ainsi que les précautions à respecter dans la fabrication des réactifs, la préparation et la purification des échantillons, font également l'objet d'une étude critique.

## Zusammenfassung **

Es wird ein Überblick gegeben über die Auswertung der Fluorescenz für Analysen von Pesticid-Rückständen und es werden neue Gebiete untersucht, auf denen diese Technik brauchbar sein dürfte. Weiter wird ein Überblick gegeben über die Vorzüge und Grenzen der Fluorescenzmethoden sowie auch über die bei der Herstellung der Reagenzien und bei der Aufarbeitung und Reinigung der Proben erforderliche Sorgfalt.

---

* Traduit par R. TRUHAUT.
** Übersetzt von G. HECHT.

128     D. MacDougall

## References

Adams, J. M., and D. MacDougall: Report No. 7075 Chemagro Corp., Kansas City, Missouri, 1961.

Anderson, C. A., J. M. Adams, and D. MacDougall: Photofluorometric method for determination of Co-Ral ($O$-$\beta$-chloro-4-methylumbelliferone $O,O$-diethyl phosphorothioate) residues in animal tissues. J. Agr. Food Chem. 7, 256 (1959).

Arison, B. H., G. V. Downing, R. A. Gray, M. A. Manzelli, J. D. Neuss, O. Speth, N. R. Tzenner, and F. J. Wolf: Abstracts of Papers, 132nd National Meeting Amer. Chem. Soc., New York City, p. 35 C (1957).

Baker, W., J. B. Harborne, and W. D. Ollis: Characterization of primary aliphatic amines by reaction with o-acetoacetylphenol and by paper chromatography. J. Chem. Soc. 1952, 3215.

Barr, C. G.: Investigations on the fluorometric determination of malic and succinic acids in apple tissue. Plant Physiol. 23, 443 (1948).

Brandt, R., and N. D. Cheronis: Fluorometric detection and determination of organic compounds. I. A rapid microsemiquantitative test for acetone and some spectrophotofluorometric data on this detection. Microchem. J. 5, 110 (1961).

Braunsberg, H., and V. H. T. James: Some observations on fluorometric determinations. Anal. Biochem. 1, 443 (1961).

Cheng, E. W., and W. Burroughs: Determination of small amounts of diethylstilbestrol in feeds. J. Assoc. Official Agr. Chemists 38, 146 (1955).

Chomse, H., and I. Arend: Microchemical detection of phthalic acid. Chem. Tech. (Berlin) 11, 377 (1959).

Cohn, V. J. Jr., and P. A. Shore: A microfluorometric method for the determination of agmatine. Anal. Biochem. 2, 237 (1961).

Conn, R. B. Jr., and R. B. Davis: Green fluorescence of guanidinium compounds with ninhydrin. Nature 183, 1053 (1959).

Druckrey, H., D. Schmöhl, and R. Preussmann: Fluorescent impurities in liquid paraffin and organic solvents. Arzneimittel-Forsch. 9, 600 (1959).

Fiegl, F.: Spot tests in organic analysis. 5th English ed. Princeton: Van Nostrand 1956.

Gehauf, B., and J. Goldenson: Detection and estimation of nerve gases by fluorescence reaction. Anal. Chem. 29, 276 (1957).

Goodyear, J. M., and N. R. Jenkinson: Irradiation fluorimetric method for estimation of diethylstilbesterol in beef liver tissue. Anal. Chem. 33, 853 (1961).

Hanker, J. S., A. Gelberg, and B. Witten: Fluorometric and colorimetric estimation of cyanide and sulfide by demasking reactions of palladium chelates. Anal. Chem. 30, 93 (1958).

Hercules, D. M., and L. B. Rogers: Fluorometric determination of 1- and 2-naphthol in mixtures. Anal. Chem. 30, 96 (1958).

Hornstein, I.: Spectrophotofluorometry for pesticide determination. J. Agr. Food Chem. 6, 32 (1958).

Hosoi, M.: A fluorometric method for the determination of estrogens in urine and plasma. Kanazawa Daigaku Kekkakm Kenkyusho Nempo 16, 145 (1958).

Ichimura, Y.: Fluorescence of coumarin derivatives. I. Fluorometric analysis of warfarin. Yakugaku Zasshi 79, 1079 (1959).

Inai, T.: Fluorometric method for the determination of human urinary estrogens. Shikoku Igaku Zasshi 16, 735 (1960).

Loeffler, W. W., and D. MacDougall: Report No. 5119 Chemagro Corp., Kansas City, Missouri, 1960.

MacDougall, D.: The use of fluorometric measurements for determination of pesticide residues. Residue reviews 1, 24 (1962).

Munsey, V. E.: Diethylstilbesterol in cattle feed. J. Assoc. Official Agr. Chemists 41, 316 (1958).

Pesez, M., and J. Ferrero: A fluorescence test in the paper chromatography of aliphatic acids. Bull. Soc. Chim. Biol. 39, 221 (1957).

Sawicki, E., and T. W. Stanley: Fluorescence spot tests for glyoxal pyruvaldehyde, salicylaldehyde and some other aromatic aldehydes. Chemist Analyst 49, 107 (1960).

SHORE, P. A., A. BURKHALTER, and V. H. COHN, JR.: A method for the fluorometric assay of histamine in tissues. J. Pharmacol. Expt. Therap. **127**, 182 (1959).

SORENSEN, L. M., and J. R. MATZKE: Stable 2-naphthol solution for fluorometric determination of malic acid. Chemist Analyst **47**, 20 (1958).

THERIAULT, R. J., W. C. FRIEDLAND, M. H. PETERSON, and J. C. SYLVESTER: Fluorometric assay for gibberellic acid. J. Agr. Food Chem. **9**, 21 (1961).

THOMMES, G. A., and E. LEININGER: Fluorometric determination of o- and m-hydroxybenzoic acids in mixtures. Anal. Chem. **30**, 1361 (1958).

— — Fluorometric determination of o-phthalic acid. Talanta **5**, 260 (1960).

WEIL-MALHERBE, H., and A. D. BONE: Chemical estimation of adrenaline-like substances in blood. Biochem. J. **51**, 311 (1952).

# Infrared and ultraviolet spectrophotometry in residue evaluations

By

ROGER C. BLINN *

With 3 figures

## Contents

## I. Introduction

In order to begin this discussion of infrared and ultraviolet spectro-photometry in residue evaluations, it is important to discuss briefly the broad field of spectrophotometry and to define the various regions of the electromagnetic spectrum of interest.

The electromagnetic spectrum is the complete system of radiant energy or, in other words, of energy propagated in wave form within the wave-length limits of $10^{-10}$ to $10^5$ cm. The portion of the electromagnetic spectrum of interest to this discussion is the region from $10^{-6}$ to $3 \times 10^{-2}$ cm. or, in more familiar terms, from 10 m$\mu$ through 300 $\mu$. This encompasses the far-ultraviolet region, which is often referred to as the vacuum-ultra-violet from 10 to 220 m$\mu$; the near-ultraviolet region from 220 to 380 m$\mu$; the visible region from 380 to 750 m$\mu$; the near-infrared region from 750 m$\mu$ to 2.5 $\mu$, the conventional infrared region from 2.5 to 25 $\mu$, and the far-infrared region from 25 to 300 $\mu$. Each of these regions will be discussed in greater detail with reference to its usefulness for residue eva-luation. Since much of this material has previously been discussed in detail, reference is made to BLINN and GUNTHER (1963).

---

* American Cyanamid Co., Agricultural Division, Princeton, New Jersey.

Spectrophotometric measurements offer several real advantages to the residue chemist. They are:

1. The radiation which is absorbed is characteristic to the material doing the absorbing, thereby offering considerable specificity to the measurement and sometimes clues as to the identity of the absorbing compound.

2. In most circumstances, the degree of absorption of radiation is directly proportional to the amount of material in the energy path of the radiation.

3. Since many substances have large absorptivity values, they can be measured at very low concentrations.

4. Spectrophotometric measurements are, for practical purposes, non-destructive; thus the absorbing material can be retrieved for further examination and evaluation.

## II. Colorimetry

While the present objective is to review the use of infrared and ultraviolet spectrophotometry in residue evaluations, certainly something should be said about the use of visible spectrophotometry for these purposes, since the quite narrow visible region is sandwiched between the much broader infrared and ultraviolet regions, and because of the widespread use and acceptance of colorimetric procedures. Colorimetric methods have the unique advantage of requiring relatively low-cost equipment. Therefore, availability of colorimeters usually is a minor problem to the residue analyst in contrast to the availability of infrared or ultraviolet spectrophotometers. However, the use of colorimetry, as popularly applied, usually results in the loss of several of those previously enumerated advantages inherent in spectrophotometric measurements. These losses of advantages result from the necessity, in most instances, for introducing a chromophoric or auxochromic group into the compound of interest or into a derived reaction product, since most pesticide or food additive compounds are not of themselves colored. Thus, colorimetric procedures are in general destructive of the parent compound of interest and little value is to be expected from further examination of the colored derivative. Also, considerable loss of specificity results during the color-producing reaction because most chromophoric or auxochromic reagents react with functional groups, and respond in varying yields to all such groups that may be present in the reaction mixture, whether their presence is due to natural products from the substrate, impurities an by-products of the compound of interest components of the formulation, or solvents and reagents used during the analytical procedure. Specificity can be gained, however, by suitable use of isolating or cleanup procedures. Because complete isolation of a single chemical compound from the varied and copious amounts of extraneous materials composing the substrate is extremely difficult to achieve, complete specificity is rarely gained.

## III. Ultraviolet spectrophotometry

The far-ultraviolet region or, as it is often called, the vacuum-ultraviolet from 10 to 220 m$\mu$ is a high-energy region where absorption is due primarily to shifts in energy levels of molecules up to ionization states and are characteristic of the atom rather than of the molecule. Since readily available instrumentation now exists only for the portion of the far-ultra-violet region above 140 m$\mu$, only this region will be discussed. There are two features that simultaneously make this region attractive and yet difficult to utilize successfully. Virtually all organic compounds absorb energy in this region and have very high absorptivities. Thus, the availability of trans-parent solvents is practically limited to water and very tediously purified paraffin hydrocarbon solvents. Also, essentially complete isolation of a residual compound would certainly be required for success. Perhaps the suggestion by KAYE (1962) for far-ultraviolet spectrophotometry coupled with vapor phase chromatography might prove useful.

It is in the realm of gases, however, that this region has to date proved most valuable. The far-ultraviolet spectra of compounds in the vapor phase can be quite characteristic and, thus, prove valuable in establishing identity. As an example, in Fig. 1 is shown the spectrum of dimethyl sulfide. Several protein hydrolysate mixtures have shown considerable attrac-tancy for the walnut husk fly. In an attempt (GUNTHER, BLINN, and BARNES, unpublished) to establish the identity of the at-tractive principle from these mixtures, the major vapor phase component in the atmosphere above the hydrolysates was col-lected both by vapor phase chro-matography and by direct con-centration, using liquid nitrogen condensation. These isolation

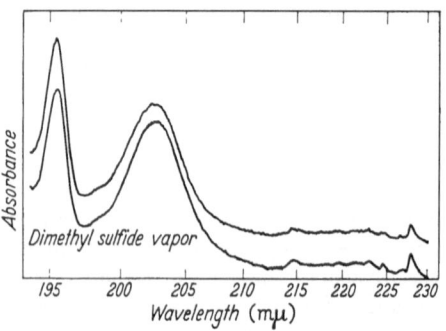

Fig. 1. The far-ultraviolet spectrum of dimethyl sulfide vapors

procedures, as well as characterizing procedures, were qualitatively and quantitatively evaluated by means of far-ultraviolet spectrophotometry due to the unique spectrum. The major gaseous component was ultimately proved conclusively to be dimethyl sulfide, with comparative far-ultraviolet spectro-photometry being the most definitive of the several instrumental evaluations used.

The near-ultraviolet region from 220 to 380 m$\mu$ is the ultraviolet region most commonly exploited for residue evaluations, partly because the availa-bility of excellent instrumentation coinciding with the development of the widespread use of modern insecticides and partly because of the wealth of transparent solvents in this region. Naturally, in order for successful utili-zation of the near-ultraviolet region for residue evaluations, one must have a compound which will absorb ultraviolet energy and have high absorpti-vity. One of the requirements for a compound to absorb near-ultraviolet energy is that it posses double bonds in its molecular structure, usually in combination and in conjugation. Thus, for a compound with little or no near-ultraviolet absorptivity, introduction of double-bonded groups, ideally in conjugation, allows its ultraviolet determination. Now one might say, "This practice of introducing chromophores is no different from that viewed critically for colorimetry." This is true, but with one difference: usually only minor conversion of a compound is required to achieve ultraviolet absorption in contrast to the strenuous measures required for its conversion to a colored compound. Thus, further instrumental scrutiny of the converted ultraviolet absorber will often be of value, whereas this is seldom true with the converted colored compound.

Because the ultraviolet absorption of most absorbing materials is so intense, measurements are invariably made upon their dilute solutions in suitable solvents. However, advantage might well be taken of any of the sampling techniques commonly used in infrared work for ultraviolet spectrophotometry. Such techniques as potassium bromide pellets, mulls, and thin-films might well take their places as valued aids. Incidentally, there has recently been reported the use of potassium bromide pellets in ultraviolet spectrophotometry (WAGGONER and CHAMBERS 1961).

## IV. Infrared spectrophotometry

The near-infrared region from 750 m$\mu$ to 2.5 $\mu$ gives spectral information which is similar to and usually complementary to that obtained in the conventional infrared region. The absorption bands observed in this region are mostly due to overtone bands from the stretching between a hydrogen atom and the atom to which it is chemically bonded, such as carbon, nitrogen, oxygen, and sulfur. The wavelengths at which these vibrations occur are quite characteristic of the functional group involved and, in contrast to the conventional-infrared region, are less affected by the structure of the remainder of the molecule. Because these bands are overtone bands and possess smaller absorptivities, energy paths must be longer than those required in the conventional-infrared region. Although excellent readily available instrumentation exists, few if any analytical applications of the near-infrared region for residue evaluations have been reported. This is undoubtedly partly due to the lower absorptivities inherent in this region, lack of specificity of the absorption bands, and to the abundance of extractable naturally occurring hydrogen-containing interfering material in foodstuffs. Use of longer energy-path cells and meticulous attention to isolative procedures should overcome some of these disadvantages.

Since very little, if any, work applicable to residue evaluations is being done in the far-infrared region beyond 25 $\mu$, there will be no discussion of this region. This leaves only the conventional-infrared region from 2.5 to 25 $\mu$, which is currently being exploited most valuably and successfully for residue work. It is in this region that identifications can most reliably be achieved, and is also well suited to quantitative work. There is readily available a wide variety of excellent instrumentation with a wide variation in cost, thus permitting even the modest budget to encompass this valuable residue laboratory tool.

Because all organic compounds absorb in this infrared region, the resultant scarcity of suitable transparent solvents led to the widespread use in the past of short energy-path cells. Because of this practice, the impression *has been* that the conventional-infrared region was suitable only for multi-milligram amounts of absorbing material and that the region is entirely unsuited for the extremely minute amounts of substances encountered in residue determinations. As is evident from Beer's law, however, sensitivity can be achieved either by increasing the energy-path length of the sample cell or by increasing the concentration through miniaturizing the volume of absorbing solution through which the energy passes. Alternatively, electronic amplification of the signal from the spectrophotometer will also result in increased sensitivity. Simultaneously with the achievement of increased sensitivity, however, there must be increased awareness of interferences due to inadequate cleanup of the substrate, to solvents and reagents, and to handling. Miniaturization of sample size can be aided by use of the readily available beam-condensing optics useful for concentrating sufficient energy through the necessarily small apertures of ultramicro cells designed to hold microliter amounts of solution or ultramicro potassium bromide pellets. The use of long energy path cells, such as the 5-mm. energy-path cavity cells, which are inexpensively commercially available, results in increased energy-path length with very little increase in volume of solution and thus, increased sensitivity. Complete infrared spectra of some 65 pesticides from 2.5 to 25 $\mu$ obtained using dilute solutions in carbon disulfide and carbon tetrachloride in 5-mm. cavity cells has recently been publsihed (BLINN and GUNTHER 1962 and 1963).

Potassium bromide pellets, in contrast to solvents, possess the advantage of being transparent throughout the conventional-infrared region. Other salts are also useful for pellet techniques, such as sodium chloride, thallous bromide, thallous chloride, and silver chloride (SMALLWOOD and HART 1963).

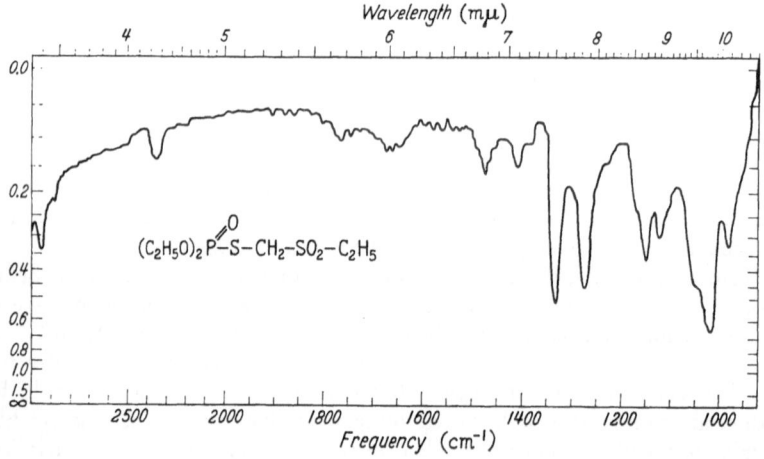

Fig. 2. Near-infrared spectrum of 7 $\mu$g. of Thimet sulfone obtained from 0.5-mm. micro-potassium bromide pellets

Considerable sensitivity can be achieved by using micro-dies and micro-pellet holders in conjunction with beam condensers. This technique certainly offers the advantage of full infrared scrutiny for identification purposes. Fig. 2 is the spectrum of the oxygen analog of Thimet sulfone, obtained from 7 $\mu$g. of the compound in a 0.5-mm. micro-pellet. Shown in Fig. 3 is the actual pellet from which this spectrum was recorded. As can be seen from this figure, only a few milligrams of potassium bromide are required and thus, as mentioned earlier, exceptional care must be exercised to avoid

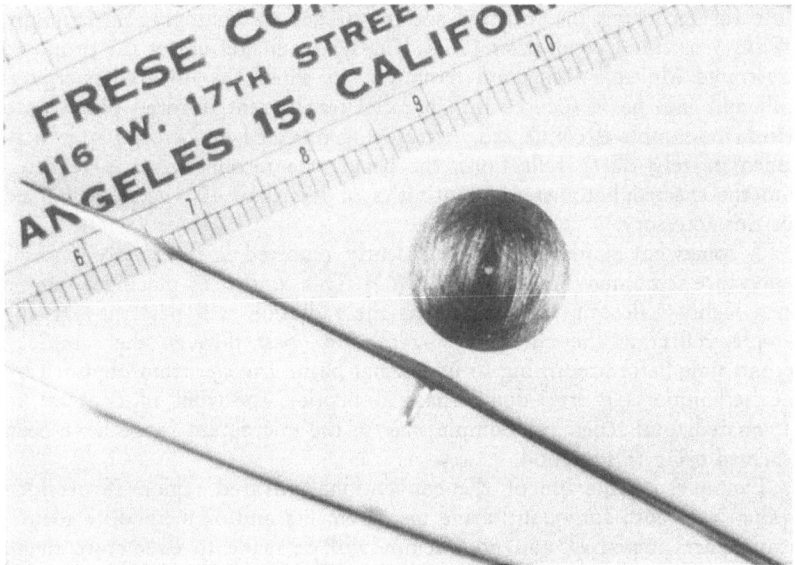

Fig. 3. The 0.5-mm. micro-potassium bromide pellet from which Fig. 2 was obtained; the pellet is contained in the hole in the stainless steel disc

interferences from solvents, handling, atmosphere, and other environmental conditions. Care in interpreting effects of solvents on spectra should also be considered, such as effects of solvents on the infrared spectra of organo-phosphorous compounds (FERRARO 1963). Quantitation of potassium bromide pellet techniques is extremely difficult, and it is advised that quantitative work be done in solutions. However, use of internal standard techniques can, in many instances, allow quantitative estimations to be accomplished. The use of mulls for the microquantities available in residue work offers few, if any, advantages over the use of potassium bromide pellets for identification purposes. However, spectra from microgram-quantities of material are achievable, as reported recently for narcotics (MILLS 1963).

There has been considerable recent interest in the use of attenuated total reflectance techniques. In these techniques, the beam of infrared energy is reflected into the spectrophotometer by means of the back face of a prism, to which the sample is in intimate contact, resulting in a spectrum as though a portion of the energy escaped a short distance from the back face of the prism into the sample before returning to the prism and into the spectrophotometer. These absorption-like spectra have two important and unique features: the band intensities are the equivalent of an extremely shallow penetration of 5 $\mu$ or less into the sample and they are independent of the thickness. Micro- and multiple reflection accessories are available commercially and are extremely useful for polymer and other high-molecular weight work. However, their usefulness for the minute quantities of material inherent in residue problems is questionable, unless it be in the field of packaging. Recently, there has been reported an accessory using this principle for recording the infrared spectra of gas chromatographic fractions (*Wilks Scientific Corporation* 1963). The system makes use of the proposed Frustrated Multiple Reflection technique, in which the infrared energy is split and each beam focused into parallel transparent infrared plates onto which the sample is condensed. After being reflected down the plates with approximately thirty reflections, the beams are recombined and reflected into the spectrophotometer. Sensitivities of less than 100 $\mu$g. are claimed for this accessory.

A somewhat similar technique recently reported is the micro specular reflectance technique (SLOANE *et al.* 1963). This consists of placing a sample on a highly reflecting surface so that the radiation will pass through the sample, reflect at the mirrored surface, then pass through the sample a second time before returning to its normal path. The spectrum obtained by this technique is a true double-pass absorption spectrum in contrast to attenuated total reflection. Sample sizes in the microgram range have been achieved using this method.

Examples of the use of the conventional-infrared region for residue evaluations, both for quantitative measurements and/or metabolite identification are numerous, and no attempt will be made to enumerate them. Mention must be made, however, of the valuable use of two of the more recent popular chromatographic techniques for isolation and cleanup purposes. This reference is, of course, to vapor phase and thin-layer chromatography, both of which lend themselves admirably to the separation, isolation, and cleanup of a substance and its metabolites. These materials can then be collected or eluted for infrared scrutiny. It must be emphasized yet again that special care must be taken to minimize contamination due to solvents and handling. It must also be emphasized again that although a cleanup procedure is successful for a material on one substrate, it must be evaluated for each other substrate of interest as the adequate cleanup of each substrate is a research problem in itself.

The advantages of the infrared region are the abundance of absorption bands which provide greater specificity of determination, possible identification of the absorbing material, and possible minimization of background absorption due to substrate and reagents by use of alternate absorption bands. Although absorptivities in the infrared region are not so great as in the ultraviolet region, adequate sensitivity is achievable by means of miniaturization, long energy-path cells, and electronic expansion of the signal from the specrtophotometer.

## V. Conclusion

In conclusion, it may be stated that when confronted with a residue problem, a careful appraisal of all of the chemical and physical properties of the substance of interest and its possible metabolites is essential for success. Appraisal of these properties will indicate whether infrared or ultraviolet spectrophotometry, or both, will be useful and possible means of achieving sufficient isolation for their successful use. As has been indicated before, "In residue chemistry, 'cleanup' is your most important problem."

## Summary

The important principles of infrared and ultraviolet spectrophotometry, as previously discussed by BLINN and GUNTHER (1963), are restated in the light of new data, information, and procedure techniques. These principles are:

1. The radiation which is absorbed is characteristic to the material doing the absorbing.

2. The degree of absorption of radiation is directly proportional to the amount of material in the energy path of the radiation.

3. Measurements at very low concentrations of a substance are often possible.

4. Spectrophotometric measurements are essentially non-destructive.

So that the advantages of the convenient vapor phase and thin-layer chromatographic procedures may be realized, techniques described in the current literature for achieving increased spectrophotometric sensitivities are discussed.

## Résumé*

Les principes fondamentaux de la spectrophotométrie infra-rouge et ultra-violette précédemment discutés par BLINN et GUNTHER (1963) sont repris à la lumière de données et de techniques nouvelles. Ces principes sont les suivants:

1. la radiation absorbée est caractéristique de la substance absorbante,

2. l'absorption de la radiation est directement proportionnelle à la quantité de matière traversée par la radiation,

3. les mesures sont souvent possibles pour des concentrations très faibles,

4. les mesures spectrophotométriques sont essentiellement non destructrices.

Afin de bien réaliser les avantages des techniques commodes de chromatographie en phase gazeuse et en couche mince, l'auteur discute les techniques décrites dans la littérature pour améliorer la sensibilité des méthodes spectrophotométriques.

## Zusammenfassung**

Die wichtigsten Prinzipien der Infrarot- und Ultraviolett-Spektrophotometrie, die früher von BLINN und GUNTHER (1963) besprochen wurden, werden im Lichte neuer Tatsachen, Informationen und Verfahrensweisen nochmals formuliert. Diese Prinzipien sind:

1. Die absorbierte Strahlung ist charakteristisch für das absorbierende Material.

---

* Traduit par R. MESTRES.
** Übersetzt von O. R. KLIMMER.

2. Der Absorptionsgrad der Strahlung ist direkt proportional zur Menge des Materials, das im Energiebereich der Strahlung liegt.

3. Oftmals sind Messungen sehr niedriger Konzentrationen eines Stoffes möglich.

4. Spektrophotometrische Messungen verursachen wesensgemäß keine Zerstörung des Untersuchungsmaterials.

Damit die Vorteile geeigneter Gas- und Dünnschichtchromatographie-Methoden klar erkannt werden können, werden in der gängigen Literatur zur Erzielung erhöhter Spektrophotometrieempfindlichkeit beschriebene Techniken besprochen.

### References

BLINN, R. C., and F. A. GUNTHER: The utilization of infrared and ultraviolet spectrophotometric procedures for assay of pesticide residues. Residue Reviews **2**, 99 (1963).
— — The promising utility of infrared assay of pesticides and their residues. I, II, and III. S.R.I. Pesticide Research Bulletin **2** (3), 1 (1962); Ibid. **2** (4), 3 (1962); Ibid. **3** (1), 5 (1963).
FERRARO, J. R.: Solvent effects on the infrared spectra of organophosphorus compounds. Appl. Spectroscopy **17**, 12 (1963).
KAYE, W. I.: Far-ultraviolet spectroscopic detection of gas chromatograph effluent. Anal. Chem. **37**, 287 (1962).
MILLS, A. L.: Infrared identification of microgram quantities of heroin hydrochloride. Anal. Chem. **35**, 416 (1963).
SLOANE, H. J., T. JOHNS, W. J. CADMAN, and W. F. ULRICH: Infrared examination of micro samples from gas chromatographic effluent. 14th Pittsburgh Conference, Anal. Chem. and Appl. Spectroscopy, Pittsburgh, Pa., March 1963.
SMALLWOOD, S. E. F., and P. B. HART: Thallous bromide as a disk material in infrared absorption spectroscopy. Spectrochim. Acta **19**, 285 (1963).
WAGGONER, W. H., and M. E. CHAMBERS: Solid state ultraviolet spectra. J. Org. Chem. **26**, 298 (1961).
*Wilks Scientific Corporation:* GC-IR analyzer. Bull. 1 (1963).

# Automatic wet chemical analysis as applied to pesticide residues

By

George D. Winter * and Andrés Ferrari *

With 8 figures

## Contents

## I. Introduction

A number of papers have been written to describe the application of some of the most modern and elegant instruments to pesticides residue analyses. Most of these instruments are based on physical mensuration and, in general, they have been employed to achieve greater sensitivity and/or speed of analysis.

The AutoAnalyzer instrumental system performs classical or modified wet chemical analyses, automatically. It is our feeling that the majority of the analyses required for pesticide residue determinations could be performed automatically, by more or less standard wet chemical procedures, employing the AutoAnalyzer.

The instrumental system itself, has been adequately described in the literature many times (Skeggs 1957, Ferrari *et al.* 1959, Lundgren 1960) and, therefore, will not be described again here.

## II. Applications

The first application of this system in the pesticide residue field, was derived from a clinical application of an acetylcholinesterase (ChE) activity determination (Winter 1960). In this method, extracts containing phosphate ester insecticides are incubated with a standard ChE solution. As

* Research Laboratories, Technicon Instruments Corporation, Chauncey, New York.

shown in Fig. 1, a ChE substrate, weakly buffered, and the insecticide sample with an appropriate diluent are aspirated simultaneously, then mixed, segmented, and fed into a 37° C. incubation coil. During the incubation period, the ChE is partially inhibited by the insecticide and the

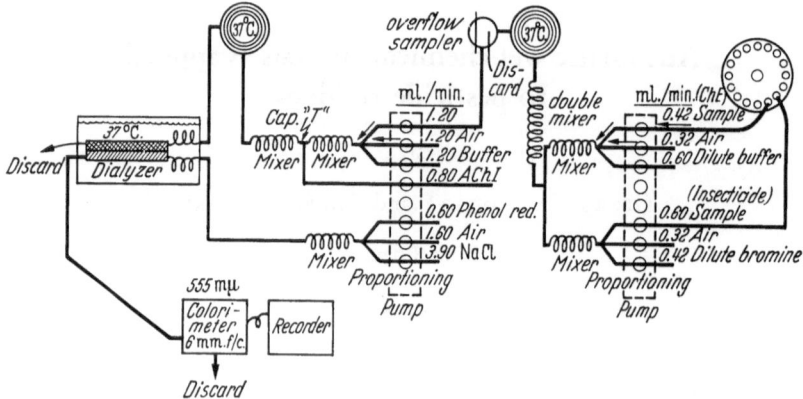

Fig. 1. Flow diagram for phosphate ester insecticides by cholinesterase inhibition

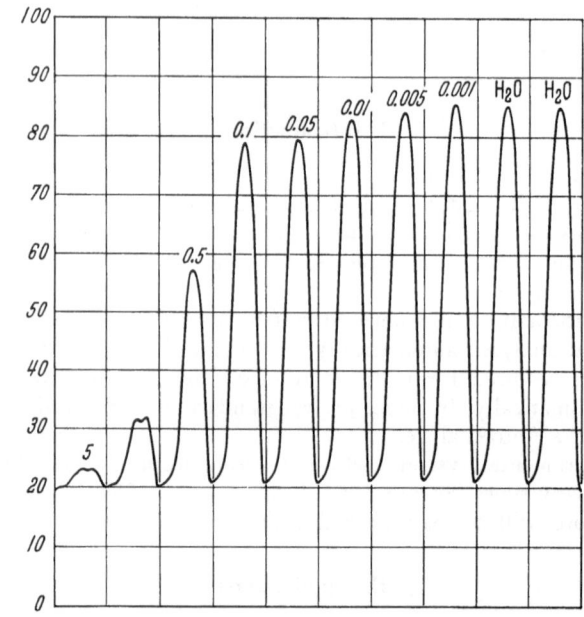

Fig. 2. Recording from inhibition of CHE by standard series of parathion solutions (numbers are μg. of parathion/ml.) with 0.04 percent bromine water added

inhibition has been found to be proportionate to the amount of phosphate ester present in the sample. A continuous aliquot of the incubation bath effluent is then buffered and mixed with acetylcholine iodide (AChI). A second incubation is then performed. The product of this incubation is acetic acid, from the hydrolysis of the acetylcholine substrate by the action of that portion of the ChE that was not originally inhibited by the insecticide. The acetic acid is proportionate to the concentration of this remaining ChE. Subsequently, continuous dialysis is employed to remove interference, proteins, etc. A solution of phenol red indicator and sodium chloride is used as the recipient for the principal product of this dialysis (acetic acid).

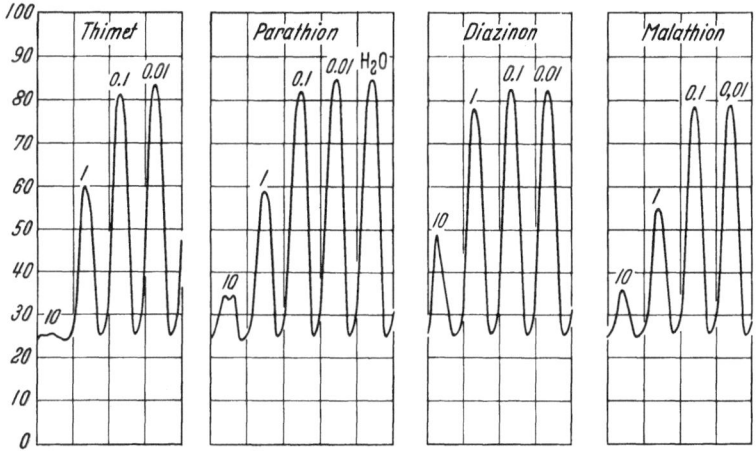

Fig. 3. Typical *in vitro* inhibition of CHE by standard solutions of four insecticides (in μg./ml., no bromine water)

The recipient stream is fed into a flow cuvette in the colorimeter and inspected at a wave length of 555 mμ. The output of the colorimeter is recorded on a strip chart recorder.

The basic reactions are: 1) partial inhibition of a standard ChE solution, present in excess; 2) hydrolysis of acetylcholine to acetic acid by the non-inhibited ChE in the system, and 3) quantitation of the acetic acid by the color change of a buffered phenol red indicator.

Fig. 2 illustrates the recordings obtained from the inhibition of ChE by a standard series of parathion solutions. These solutions ranged from 5 μg to 0.001 μg of parathion per ml. This sensitivity was achieved by the addition of a dilute bromine solution to the system to oxidize the parathion and increase its *in vitro* activity.

Fig. 3, illustrates typical *in vitro* inhibition of ChE by standard solutions of four different insecticides.

Most chemical residue analyses techniques require ashing, wet digestion, or distillation, for isolation and/or purification of the residue material. The development of a continuous digestion module for the AutoAnalyzer system has made it possible to carry out these procedures and thus extend the scope of the system, in this and many other fields (Ferrari 1960).

Fig. 4 illustrates the module for digestion, with the remainder of the modules required for common digestion techniques.

Fig. 5 shows a diagrammatic cross-section of the digestion glass helix of Fig. 4 at the input and output ends of the system. The combined reagent

Fig. 4. Digestion and associated modules. Right to left: sampler, proportioning pump, digestor, proportioning pump, colorimeter, and recorder (top)

and sample stream are fed into the primary circular groove of the digestion helix. Rotation of the glass helix at a constant speed continuously conveys the mixture of digestion reagent and sample through the furnace of the

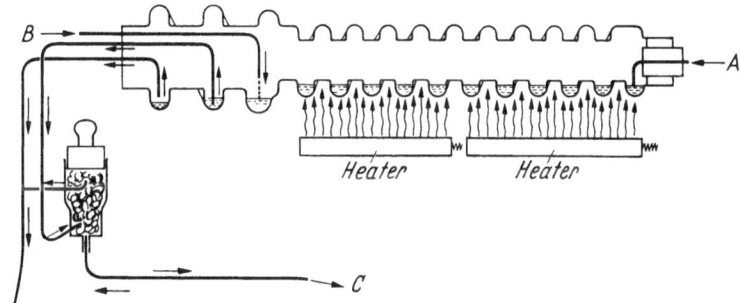

Fig. 5. Diagrammatic cross-section of the digestion glass helix of Fig. 4. *A*, sample and digestion fluid input; *B*, diluent input; *C*, sample to analytical system

digestor. Residence time in the digestor may be changed by varying the r.p.m. of the helix. Temperatures of the three separate heating zones of the digestor are made adjustable by the use of individual variacs.

At the exit end of the digestor helix a constant stream of distilled water for dilution is introduced into the first circular groove. The diluted sample is then fed into a second circular groove, from which a pipette, in contact with the bottom of the circular groove, aspirates the total fluid input. A third pipette, located in the third and last circular groove, is included as a safety measure, in the event that there should be an overflow of digestion fluid beyond the middle groove. The combined diluent and digestion fluid are forcibly aspirated into a glass mixing chamber where inflow of air during the period between samples agitates and mixes the contents. From this mixing chamber a suitable aliquot is proportioned into the analytical system, while the remainder is constantly aspirated into a waste bottle.

Fig. 6 illustrates the flow diagram for the automation of wet digestion techniques. The sample is introduced into a stream of the digesting fluid previously heated to 95° C. by passage through a constant temperature bath. After digestion and suitable dilution, the combined fluids are aspirated into the glass mixing chamber, and a suitable aliquot is introduced into the analytical system.

A stream of 30 percent sodium hydroxide with 4 percent EDTA added is employed to neutralize the excess acid from the digestion. The near neutral solution of digested sample is mixed with a stream of ammonium molybdate, and then introduced into a stream of aminonaphthol sulfonic acid. The entire stream is pumped through a 95° C. bath for the development of the molybdenum blue complex as a function of phosphate concentration.

This basic methodology has been modified by a group headed by M. E. GETZ of the U. S. Food and Drug Administration.[1]

[1] It is our understanding that the method is currently under test and that the results to date have been very satisfactory (GETZ 1963).

The digestor module can be modified to include distillation after digestion of the organic material. The specimen is introduced into the digestion helix with the digestion fluid. During the first heating stages, destruction

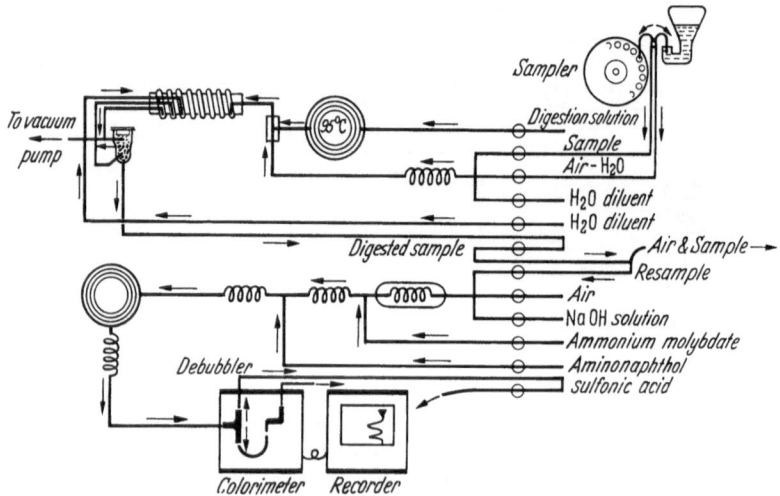

Fig. 6. Flow diagram for automation of wet digestion techniques

of the organic matter occurs, and further down the helix the volatile constituents are carried off in the vapor phase, aspirated under vacuum through a collection tube, and sent through a condenser.[2] The condensate is then fed into the analytical system for quantitation. Successive samples are separated in the system by a wash of digestion fluid and water vapor that flows continuously, while the samples are introduced intermittently.

Fig. 7 illustrates a cross-section of the digestion and distillation helix (E). Samples and digestion fluid are introduced into the circular groove at the extreme right. Aliquots are distributed to the helices which convey the fluid through the furnace. In this method, the helix is revolved at high speed to effect rapid transfer of the sample through the digestion and distillation phases, and to provide adequate wash and separation between specimens. Digestion, as evidenced by charring and subsequent clearing, occurs in the first portion of the system. The vapor (D) evolved in the digestion phase is aspirated further downstream and drawn by vacuum into a flared-end glass tube which is located immediately beyond the last heating stage (2) of the furnace. The vapor is then fed through a jacketed condensing coil (F) and the condensate continues to a sample mixing chamber (H) which provides through mixing by a bubbling process. An aliquot of the condensate is continuously aspirated into the analytical system (J).

---

[2] This condenser may be used straight, or packed with glass beads to increase surface area.

Fig. 8 is a diagrammatic illustration of the fluid routing through the digestion-distillation and analytical systems.

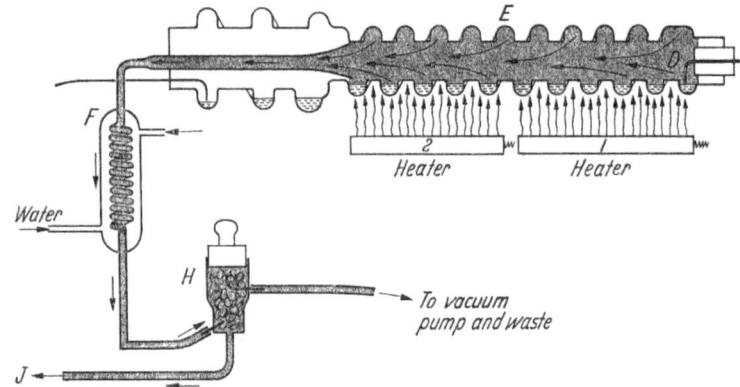

Fig. 7. Diagrammatic cross-section of the digestion and distillation helix: *D*, vapor; *E*, helix; *F*, condenser; *H*, sample mixing chamber; *J*, condensate to analytical system

Workers at the Boyce Thompson Institute for Plant Research (WEINSTEIN and MANDL 1963) have developed this technique and applied it to a

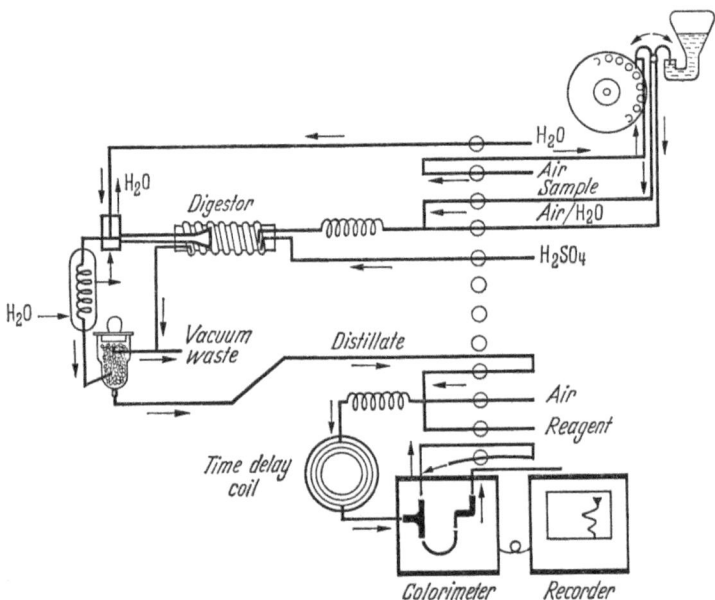

Fig. 8. Flow diagram of fluid routing through the digestion-distillation and analytical systems

fluorine analysis system. They have been able to determine fluorine in plant tissue at concentrations smaller than 0.1 $\mu$g./ml., using an alizarin complexone reaction.

## III. Discussion

This paper is being presented to demonstrate a variety of techniques that have been applied to the automation of chemical analysis in the field of pesticide residue research.

Automation of existing methods offers several advantages to the laboratory:

1. Manipulation is greatly reduced, permitting more efficient utilization of specialized manpower.

2. More samples may be processed in a given time. This means that more frequent sampling may be employed and better statistical data will result.

3. Values obtained and sample data are permanently recorded for later reference.

4. The basic methods are those of classical wet chemical procedures. The chemist is given an assurance that the procedures and therefore, the derived data, are valid.

The laborious preparation, homogenization, and extraction of the residues from bulky plant samples have not yet been automated to our knowledge. Work on this phase of the problem is currently in process, in an attempt to improve further the efficiency of our system in this application.

## Summary

Adaptations of the AutoAnalyzer analysis system have resulted in automation of some of the analyses required for residue determinations in the past.

The recent development of a continuous digestor and/or distillation module now permits (a) automation of some analyses not hitherto possible, and (b) automation of more stages of some of the other determinations, e. g., sample degradation is now possible.

Many specific residue analyses are now possible, in addition to broad group analyses for mass screening.

## Résumé *

Des adaptations du système d'analyse par l'Autoanalyseur ont conduit à l'automation de certaines des analyses imposées jadis pour les dosages de résidus.

L'élaboration récente d'un digesteur continu et (ou) d'un module de distillation permet, à présent, (a) l'automation de certaines analyses non encore réalisables jusqu'ici, et (b) l'automation de plusieurs stades dans certains des autres dosages; à titre d'exemple, la dégradation de l'échantillon est actuellement possible.

Beaucoup d'analyses spécifiques de résidus sont maintenant réalisables en plus des grands groupes d'analyses pour la discrimination globale.

## Zusammenfassung **

Die Anwendung des „Auto-Analyzer"-Verfahrens gestattet es, einige Analysen zur Rückstandsbestimmung zu automatisieren. Die neuere Ent-

---

* Traduit par S. Dormal van den Bruel.
** Übersetzt von S. W. Souci.

wicklung eines kontinuierlichen Aufschluß- und/oder Destillationsverfahrens ermöglicht a) Analysen zu automatisieren, bei denen dies bisher nicht möglich war, und b) mehrere Stufen einiger anderer Bestimmungen zu automatisieren. Dadurch ist z. B. eine Verringerung der Probemenge möglich. Neben umfangreichen Gruppenanalysen für Massenuntersuchungen sind durch das neue Verfahren viele spezielle Rückstandsanalysen möglich.

## References

FERRARI, A.: Nitrogen determination by a contiuous digestion and analysis system N. Y. Acad. Sci. 87, 792 (1960).
—, F. M. RUSSO-ALESI, and J. M. KELLY: A completely automated system for the chemical determination of streptomycin and penicillin in fermentation media. Anal. Chem. 31, 1710 (1959).
GETZ, M. E.: Personal communication (1963).
LUNDGREN, D. P.: Methods development for phosphate analysis with the Autoanalyzer. N. Y. Acad. Sci. 87, 904 (1960).
SKEGGS, L. T. JR.: Automatic method for colorimetric analysis. Amer. J. Clin. Path. 28, 311 (1957).
WEINSTEIN, L., and R. MANDL: Personal communication (1963).
WINTER, G. D.: Cholinesterase activity determination in an automated analysis system. N. Y. Acad. Sci. 87, 629 (1960).

# Determination of pesticide residues
# by neutron-activation analysis

By

V. P. Guinn * and R. A. Schmitt *

## Contents

## I. Introduction

Although activation analysis was first employed during the 1930's, it remained as a seldom-used technique until the advent of the nuclear reactor — with its attendant high neutron fluxes. It became considerably better known during the 1950's, when the activation analysis service of the Oak Ridge National Laboratory made the technique available, at least indirectly, to any interested persons. More recently, the development of lower-cost, high-performance research reactors, low-cost neutron generators (deuteron accelerators), and versatile multichannel pulse-height analyzers has brought the technique directly into many laboratories. Neutron-activation analysis has now been applied in a great many different fields of

---

* General Dynamics Corporation/General Atomic Division, San Diego, California.

science and industry, for example, the petroleum, chemical, and plastics industries, metallurgy, biology and medicine, criminalistics, and, most recently, pesticide residue studies.

Activation analysis may be defined as a method of elemental analysis in which various elements in samples are made radioactive by inducing in them nuclear reactions, followed by quantitative determination of the amounts of such induced radioactivities. The nuclear reactions may be brought about by exposing the samples to various particles: protons ($p$), deuterons ($d$), alpha ($a$) particles, electrons ($e$ or $\beta$), photons ($x$ or $\gamma$), or neutrons ($n$). The heavy positively-charged particles ($p$, $d$, $a$) have limited usefulness because of their low penetration into samples (a few microns or tens of microns) and the high cost of the accelerators needed to accelerate them through the high potentials required (several million volts). Electrons penetrate somewhat better, but suffer from very low reaction cross sections and the high cost of suitable high-energy electron accelerators. Photons ($x$-rays and gamma rays) penetrate well, but also exhibit very low reaction cross sections and require expensive high energy accelerators for their generation.

By far the most useful particle for activation analysis, to date, has been the neutron. When neutrons are slowed down to thermal energies, they exhibit relatively large reaction cross sections for capture by the nuclei of most elements. They penetrate most kinds of samples quite well and can be generated at fair intensities (fluxes) by means of relatively inexpensive accelerators, and at much higher fluxes by means of nuclear reactors. For most elements, thermal neutrons are captured with appreciable reaction cross sections to produce radioactive isotopes of the same elements that captured them. For some elements, however, thermal neutron reactions do not produce usable radioactivities, but most of these elements can be made suitably radioactive by bombardment with high energy neutrons.

Two main approaches are used in neutron-activation analysis: destructive and nondestructive. In the older, destructive, method, samples are activated with neutrons, then separated chemically into different element groups which are then counted either by beta-particle counting or gamma-ray counting. The newer, nondestructive, method involves only activation and gamma-ray spectrometry. It is much more rapid. The two methods each have their place, and their respective advantages and disadvantages. They are discussed in more detail in later sections.

## II. The theory of neutron-activation analysis

By far the most widely used form of activation analysis is that which employs "thermal" neutrons, that is, neutrons which have been slowed down until they are in kinetic energy equilibrium with their surroundings. Thus, at normal room temperature (20° C.), the average kinetic energy of thermal neutrons is 0.025 electron volts (ev) [1]. Such slow-moving neutrons

---

[1] An electron volt is the kinetic energy acquired by an electron in falling through a potential drop of one volt.

are rather easily captured by many of the stable isotopes of many of the elements. For example, when aluminum is exposed to a flux of thermal neutrons, this nuclear reaction occurs readily:

$$Al^{27} + n^1 \rightarrow Al^{28} + \gamma.$$

Ordinary stable aluminum consists entirely of the isotope of mass 27 (13 protons, plus 14 neutrons, in the nucleus). When such a nucleus captures a neutron, it remains aluminum since the atomic number, or nuclear charge (the number of protons) does not change, but increases in mass by one unit. The product nucleus in neutron-capture reactions is usually formed in an excited state which promptly drops to the ground state with the emission of one or more prompt "capture" gamma rays, indicated by the $\gamma$ in the above equation. Such a reaction is usually written in a short-hand form:

$$Al^{27}(n, \gamma)Al^{28}.$$

### a) Radioactive decay

The reaction product in this case, $Al^{28}$, is radioactive. It decays with a half life of 2.27 minutes by beta particle emission, forming stable $Si^{28}$. However, it forms $Si^{28}$ in a nuclear excited state (analogous to the electronic excited states of atoms and molecules), which promptly drops to the ground state of $Si^{28}$ with the emission of a characteristic 1.78-Mev [2] gamma ray photon. The process of radioactive decay is often shown schematically in the form of an energy-level diagram, which for the case of $Al^{28}$ is as follows:

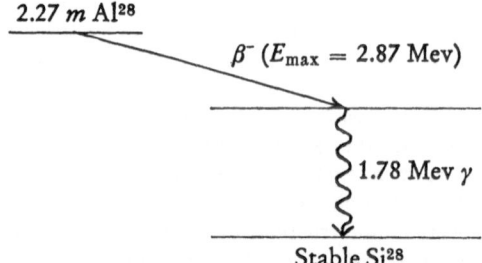

2.27 $m$ $Al^{28}$

$\beta^-$ ($E_{max} = 2.87$ Mev)

1.78 Mev $\gamma$

Stable $Si^{28}$

In the great majority of cases where neutron capture produces a radioisotope, the latter decays by negative beta particle emission, as in the example of $Al^{28}$. This is due to the fact that the original stable nucleus became neutron-rich unstable by capturing a neutron. It seeks to regain stability by getting rid of the excess neutron. Since there is not enough energy available to simply eject it, the nucleus converts a neutron to a proton, within the nucleus. This process results in the formation of a beta particle (negative) and an antineutrino ($\tilde{\nu}$), each of which escape from the nucleus, leaving behind the newly formed proton:

$$n \rightarrow p^+ + \beta^- + \tilde{\nu}.$$

---

[2] Mev is the abbreviation for the energy term or unit, million electron volts.

In some cases, however, neutron capture forms a radioisotope that behaves as a neutron-poor isotope. These decay by conversion of a nuclear proton to a nuclear neutron by one or the other of two processes, positron $(\beta^+)$ emission or orbital electron capture *(EC)*, usually from the atomic $K$ shell of electrons:

$$p^+ \to n + \beta^+ + \nu$$
$$p^+ + e^- \to n + \nu.$$

A positron is simply a positively charged electron. Positron emission is accompanied by emission of a neutrino $(\nu)$. Unless the nuclear energy available in the decay exceeds 1.02 Mev, positron emission cannot occur (but electron capture can); with available energies above 1.02 Mev both processes can occur. Illustrative cases, showing energy level diagrams for these two types of radioactive decay, are those of 12.8-hour $Cu^{64}$ and 2.6-year $Fe^{55}$:

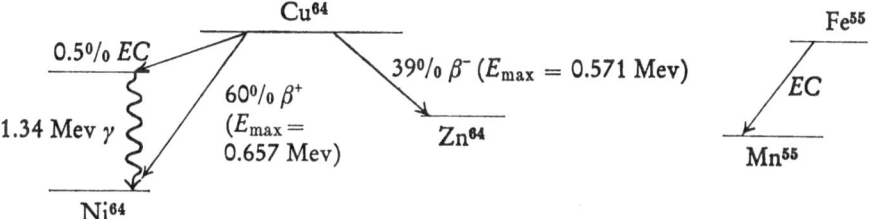

In counting operations, the neutrinoes and antineutrinoes are not detected. Emitted positrons normally slow down within the sample and undergo an "annihilation" reaction with an electron, to form two 0.51 Mev gamma ray photons going in opposite directions:

$$\beta^+ + e^- \to \gamma + \gamma.$$

In electron capture decays the resulting nucleus emits x-rays, which can be detected.

### b) Mathematical relationships

When a sample is exposed to a flux $(f)$ of thermal neutrons, the rate of formation of any particular radioisotope is given by the equation

$$Rate\ of\ formation \text{ (nuclei per second)} = Nf\sigma, \qquad (1)$$

in which $N$ is the number of target nuclei, of the appropriate type, present in the sample, $f$ is the thermal neutron flux (in neutrons/cm.² second) to which the sample is exposed, and $\sigma$ is the neutron capture cross section of the target nuclei, in cm.²/nucleus. Tables of capture cross sections usually list the cross sections in "barns". One barn is defined as $10^{-24}$ cm.²/nucleus.

Once formed, of course, any given radioactive nucleus may disintegrate, by radioactive decay, at any statistically random time. For a large number of such nuclei, the probable lifetime before decay (as opposed to the average lifetime) is what is called the "half life" of the isotope, $t_{0.5}$. It is the time required, on the average, for half of the initially present nuclei, $(N_o)$ to disintegrate. Thus,

$$N = N_0\, e^{-0.693t/t_{0.5}}. \tag{2}$$

If one activates a sample for a period of time quite long, relative to the half life of the particular radioisotope being formed, a steady state will be reached, in which the rate of formation of the radioisotope will be equal to its rate of decay. Continued irradiation then no longer increases the activity level of that particular radioisotope. This level of activity is usually termed the "saturation" level of activity. At shorter irradiation times the activity level of a given induced activity is of course intermediate between zero and the saturation level, and is given mathematically by the equation

$$A_0 = Nf\sigma\,(1 - e^{-0.693t_i/t_{0.5}}) \tag{3}$$

in which $A_0$ is the activity level of that particular radioisotope, expressed in disintegrations per second (dps), just at the conclusion of the irradiation, and $t_i$ is the duration of the irradiation, expressed in the same units of time as the half life. At very long irradiation times, where $t_i \gg t_{0.5}$, $e^{-0.693t_i/t_{0.5}}$ approaches zero, and hence

$$A_0\ (\text{saturation}) = Nf\sigma\ (\text{for } t_i \gg t_{0.5}). \tag{4}$$

At $t_i/t_{0.5}$ values of 1, 2, 3, 4, ..., the parenthetical expression of equation 3 has numerical values of 1/2, 3/4, 7/8, 15/16, ..., in other words, the expression asymptotically approaches a value of unity. The variation of the quantity, $1 - e^{-0.693t_i/t_{0.5}}$, with $t_i/t_{0.5}$, is shown graphically in Fig. 1. Because of this inherent relationship, it is rather pointless, in practice, to activate a sample for a period of time more than one or two half lives of the radio-isotope of interest.

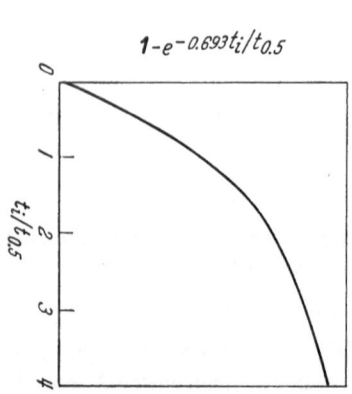

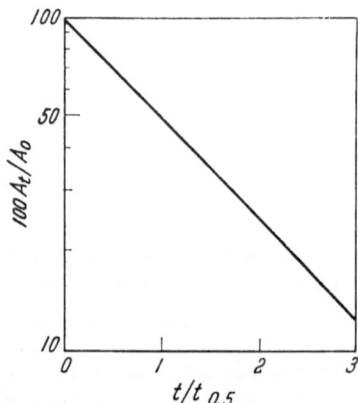

Fig. 1. Graph of $(1 - e^{-0.693t_i/t_{0.5}})$ vs. $t_i/t_{0.5}$      Fig. 2. Graph of log $A_t/A_0$ vs. $t/t_{0.5}$

After completion of the activation, each induced radioisotope decays exponentially, at a rate dependent upon its half life, following the equation

$$A_t = A_0 e^{-0.693t/t_{0.5}} \tag{5}$$

in which $t$ is the length of time the sample has decayed since $t_0$, and $A_t$ is the activity at decay time $t$. The units of $t$ and $t_{0.5}$ must be the same, since the exponent must be dimensionless. If the logarithm (base 10) of both sides of equation 5 is taken, the following expression is obtained:

$$\log A_t = \log A_0 - 0.301 \, t/t_{0.5} \, . \qquad (6)$$

Thus, a graph of the logarithm of $A_t$ versus $t$ is a straight line for each single radioactive species. The slope of the line is equal to $-0.301/t_{0.5}$, and hence provides a determination of the half life of the isotope, if the sample is counted at several different decay times. The intercept at time zero provides a measure of $A_0$. Equation 6 may also be rewritten as

$$\log (A_t/A_0) = -0.301 \, (t/t_{0.5}) \, . \qquad (7)$$

In this equation, $A_t/A_0$ is the relative activity at time $t$, compared to that at time zero. A graph of $\log A_t/A_0$ versus $t/t_{0.5}$ is a straight line which is the same for all radioisotopes. This graph is shown in Fig. 2.

### III. Counting techniques

After activation, the induced activities are measured by counting either the beta particles emitted by the radioactive sample, or the gamma rays emitted.

#### a) Beta counting

In the $Al^{28}$ example cited earlier, it was noted that $Al^{28}$ decays by beta particle emission $(\beta^-)$, promptly followed by emission of a gamma ray photon. The energy released in the beta particle emission process, in the case of $Al^{28}$, is 2.87 Mev, Although this amount of energy is released each time an $Al^{28}$ nucleus disintegrates, the beta particle may receive anything from none to all of the energy, the simultaneously emitted antineutrino receiving the rest. In tabulations of the beta particle energies of beta-emitting isotopes, the maximum energy ($E_{max}$) is listed. Thus,

$$E_{\beta^-} + E_{\bar{\nu}} = E_{max} \qquad (8)$$

for each disintegration. Since only the beta particle is detected, the observed beta particle energy distribution, even from a single radioisotope species such as $Al^{28}$, forms a continuous spectrum. A typical beta particle energy spectrum is shown in Fig. 3. The average beta particle energy is about 1/3 of $E_{max}$.

It is evident, then, that if an activated sample contains a number of

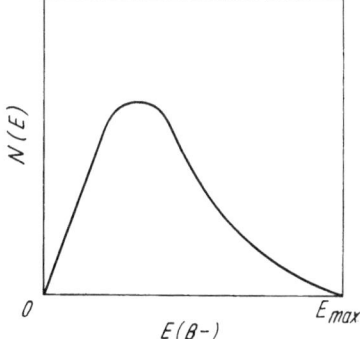

Fig. 3. Graph of typical $\beta$ energy spectrum

different induced beta-emitting radioisotopes, beta particle counting cannot readily distinguish which isotopes are producing the beta particles, or how many are due to each, since each species is emitting a whole spectrum (continuous) of beta particle energies, ranging from zero to its particular $E_{max}$. The situation is further complicated by the fact that many radioisotopes emit beta particles of two, or three, or even more different $E_{max}$ values. A good example is that of 9.5-minute $Mg^{27}$:

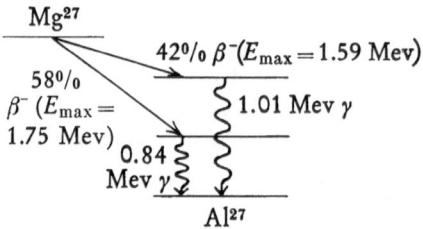

This isotope emits beta particles with $E_{max}$ values of 1.59 Mev in 42 percent of its disintegrations, and of 1.75 Mev in the other 58 percent.

As will be shown later, the gamma rays emitted by a radioisotope are monoenergetic, rather than covering a continuous energy spectrum. With a suitable counter, gamma rays of one energy (from one radioisotope) can be distinguished from gamma rays of other energies (from other radioisotopes present), and even complex mixtures can be resolved by gamma-ray spectrometry. Thus, in general, gamma-ray counting is preferred to beta-particle counting.

Unfortunately, some induced radioactivities emit only beta particles — with no gamma-ray emission. Examples of importance in neutron activation analysis are 87-day $S^{35}$ and 14.5-day $P^{32}$, to mention only a few. In such cases one is forced to do beta counting. Any interfering beta-emitters also present must then be (1) eliminated by decay, (2) corrected for by resolution of decay curves or beta absorption curves, or (3) eliminated by chemical separations.

In order to inter-compare the beta counting rates of various samples with a standard sample of the same radioisotope, it is important that self-absorption of beta particles within the samples be rigorously standardized, or else normalized by means of a calibration curve. Often, either "infinitely thick" samples (ones sufficiently thick that only beta particles emitted in the upper part of the sample can possibly emerge from the sample and reach the counter), or "infinitely thin" samples (ones so thin that self-absorption is negligible), are employed. For intermediate sample thicknesses, one must prepare a careful calibration curve, for that radioisotope in the given matrix, of relative counting rate (at the counting geometry selected) versus sample thickness for a sample of constant "specific activity" (disintegrations per minute, per gram of sample). Beta counting is usually carried out with a lead-shielded thin-window Geiger-Mueller counter, with the sample placed close to the counter window, at a carefully fixed distance.

Positrons ($\beta^+$) behave just like negative beta particles ($\beta^-$) in their self-absorption properties. However, positrons that slow down to thermal velocities are promptly "annihilated" by electrons in the sample, thereby producing annihilation (0.51-Mev) $\gamma$ photons. Hence, $\beta^+$ emitters are normally counted just like $\gamma$-emitters, rather than as $\beta$-emitters.

## b) Gamma-ray spectrometry

Since the emphasis in this paper is upon the purely instrumental (rapid, nondestructive) form of neutron activation analysis, the details and problems of beta particle counting will not be presented in any further detail.

The counting of gamma-ray photons emitted by the activated sample, rather than beta particles, is preferable for several reasons: (1) the emitted gamma rays are monoenergetic, rather than having continuous energy spectra, (2) scintillation counters are available that produce output electrical pulses that are directly proportional to the gamma-ray energy absorbed by the detector, and (3) gamma rays are so penetrating that self-absorption within the samples is usually negligible.

The standard counter for gamma-ray spectrometry work is the NaI(Tl) scintillation counter. It consists usually of a cylindrical single crystal of thallium activated sodium iodide, coupled to the photocathode face of a photomultiplier tube (PMT). Typical crystal sizes are two inches in diameter by two inches high, or three inches $\times$ three inches. The crystal is "canned" in an aluminum case to protect it from light and from attack by atmospheric moisture. A white diffuse reflector, such as MgO or $Al_2O_3$, is deposited between the crystal and the aluminum case. The Al and reflector material also absorb most of the undesired beta particles that might otherwise reach the crystal from the sample. The flat end of the crystal next to the photomultiplier tube has a glass plate sealed to it, instead of Al or reflector. A photograph of the two main parts of a NaI(Tl) scintillation counter (crystal and PMT, exploded view) is shown in Fig. 4.

When a gamma-ray photon from the sample strikes the NaI(Tl) crystal, it penetrates and then may (1) pass all the way through without interaction, (2) interact with an electron somewhere in the crystal by Compton scattering, losing only part of its energy, or (3) be completely absorbed by a photoelectric absorption or pair-production process.

In the Compton type of interaction, the $\gamma$ undergoes a sort of "glancing collision" with an electron. As a result, the $\gamma$ gives a fraction of its energy to the electron and then moves on with reduced energy, and in a different direction:

Thus, $E_o = E' + E_{e^-}$. In Compton interactions, the original $\gamma$ can lose anywhere from almost none of its energy up to a maximum value corresponding to the "Compton edge" $(E_{ce})$:

$$\frac{E_o}{1 + \dfrac{0.51}{2\,E_o}} = E_{ce}. \tag{9}$$

If a $\gamma$ interacts by photoelectric absorption, it gives all of its energy to an electron, and the $\gamma$ photon ceases to exist. If the $\gamma$ interacts by pair production, it converts part (1.02 Mev) of its energy to mass, forming a

$\beta^-$ and a $\beta^+$, and imparting the rest of its energy to these newly-formed particles as kinetic energy. This process is a good example of Einstein energy to mass conversion ($E = mc^2$). It obviously can only occur with $\gamma$ photons having energies greater than 1.02 Mev (the "rest mass" of the $\beta^-$ and the $\beta^+$ each being 0.51 Mev, in energy units).

Fig. 4. Exploded view photograph of a NaI(Tl) scintillation counter

Whenever an electron in the NaI(Tl) crystal receives energy by a Compton scattering or photoelectric absorption event, or whenever an energetic $\beta^-/\beta^+$ pair is formed by pair production, these energetic electrons move only a short distance, of the order of a millimeter or less, in the crystal before they are slowed down to almost zero velocities. Thus, their kinetic energies are imparted to the crystal. By a complex mechanism, the NaI(Tl) crystal emits about 10 percent of this absorbed energy in the form of light. This is emitted as a burst of light photons (each of about 3 ev energy), originating close to the point of the original $\gamma$ ray interaction, isotropically in all directions. The number of 3 ev photons emitted is directly proportional to the energy absorbed by the crystal in that interaction. Thus, if a 1-Mev $\gamma$ were completely absorbed, the number of 3 ev photons produced would be about 10 percent of $10^6$ ev, divided by 3 ev, or about 33,000. By reflections from the reflecting walls of the canned crystal, most of these 33,000 would fall virtually instantaneously on the photocathode of the PMT.

A typical PMT has a photocathode efficiency of about 10 percent for NaI(Tl) light emission. Thus, the 33,000 light photons cited in the example above, if all hit the photocathode, would generate about 3,300 photoelectrons within the PMT. The latter vould electrically focus these on the first "dynode", which, by secondary emission, might multiply these by a factor of perhaps three. These would be electrostatically focussed onto the second dynode, where again a multiplication of about three would occur, and so on down a chain of ten dynodes. Thus, the final electron output of the PMT, from the absorption of a single 1-Mev $\gamma$ photon, might be $3{,}300 \times (3)^{10}$, or almost $2 \times 10^8$. The size of the output pulse, for a given crystal, PMT, and PMT applied voltage, is directly proportional to the $\gamma$ ray energy absorbed by the crystal in any given interaction. At the PMT output, each burst of electrons is converted to a voltage pulse, which is then amplified further in an electronic amplifier, and then fed to a pulse-height analyzer which measures its size and stores this information.

For NaI(Tl) the relative probabilities, or cross sections ($\sigma$), of $\gamma$ interaction by Compton scattering ($\sigma_c$), photoelectric absorption ($\sigma_{pe}$), and pair production ($\sigma_{pp}$) are such that photoelectric absorption is predominant for $\gamma$ energies less than 0.3 Mev, Compton scattering is predominant for $\gamma$ energies from 0.3 Mev to 6 Mev, and pair production is the predominant mode above 6 Mev. Most of the gamma rays encountered in activation analysis are in the 0.1-Mev to 3-Mev range. Thus, even when gamma rays of only a single energy strike the NaI(Tl) crystal, a spectrum of output pulses is obtained. The typical shape of such a monoenergetic $\gamma$ ray pulse-height spectrum is shown in Fig. 5, which shows the typical appearance of the "Compton continuum", "Compton edge", and "photopeak" regions of such spectra. The location of the Compton edge is defined by equation 9. The theoretical gap between the Compton edge and the total absorption peak ("photopeak") is partially filled in, and rounded, and the photopeak is broadened from the theoretical vertical spike to a gaussian shape of quite finite width because of the statistical nature of the interaction produced in the crystal and PMT. A very good crystal-PMT combination typically shows a resolution (photopeak width at half height) of about 5.5 percent for 1-Mev $\gamma$ rays. The percent resolution varies approximately as $E_\gamma^{-1/2}$, thus being poorer (wider) at lower energies and better (narrower) at higher energies. Thus, at 0.5 Mev the resolution would be about eight percent, and at two Mev would be about four percent.

When the radioactive sample being counted is emitting gamma rays of several different energies, at various emission rates, the observed pulse-

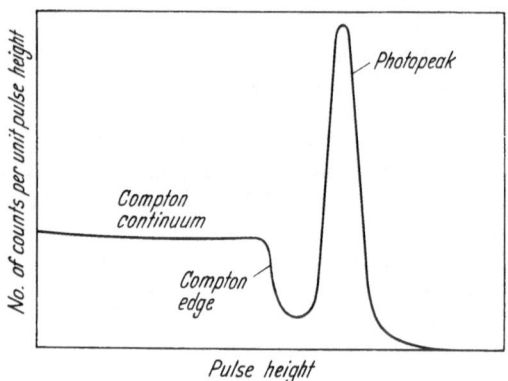

Fig. 5. Typical monoenergetic gamma ray pulse-height spectrum from a NaI(Tl) scintillation counter

height spectrum is of course more complex. However, such spectra are simply the summation of the various single-component spectra, and hence show the various characteristic photopeaks superimposed on a higher and broader Compton continuum. Examples of such complex spectra are shown later. Other features of typical pulse-height spectra, such as lead x-rays peaks (produced by $\gamma$ interactions in the lead shield), backscatter peaks (produced by 180°-scattering of $\gamma$ rays in the lead shield), iodine escape peaks (only of importance with $\gamma$ rays of very low energy), and annihilation escape peaks (of importance only for $\gamma$ rays with energies appreciably above 1.02 Mev), cannot be treated here, in the interest of brevity. One note on the role of NaI(Tl) crystal size is, however, in order. Larger crystals produce spectra with more counts in the photopeaks, as is desired, and fewer in the Compton region, because initially Compton-scattered $\gamma$ rays may be absorbed before they can escape the crystal, if the crystal is larger. This desired effect of larger size is partially offset by three factors: (1) increasing cost of larger crystals, (2) poorer resolution of larger crystals, and (3) increasing background counting rate of larger crystals. In general, a three-inch×three-inch crystal is considered a good compromise. Even so, such a crystal costs about 800 dollars (compared with 300 dollars for a 2×2 and 2,000 dollars for a 5×5).

Although earlier scintillation $\gamma$ ray spectrometry was carried out with single-channel pulse-height analyzers, modern work is done almost entirely, and easier and faster, with multichannel pulse-height analyzers (PHA). Hence, this discussion will be limited to multichannel analyzers. The modern multichannel PHA is a transistorized instrument that measures the voltage amplitude of each incoming electrical pulse from the scintillation counter (after suitable amplification in the amplifier) by a process of "amplitude-to-time-conversion", and then stores this information in a magnetic-core storage or memory. A photograph of a representative multichannel PHA is shown in Fig. 6. As the sample is being counted, each analyzed pulse appears as a flash of light on the oscillocope tube. This display shows the

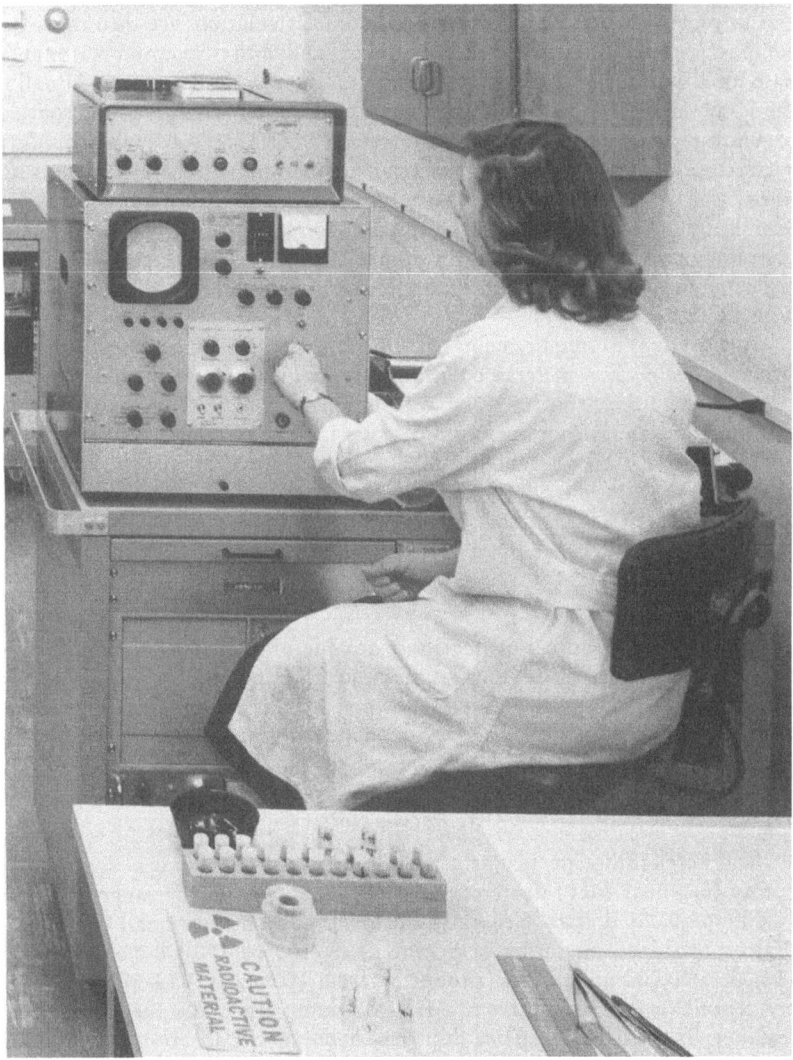

Fig. 6. Photograph of a representative multichannel pulse-height analyzer

accumulated number of pulses of a given size range (thus, in a given storage channel) as the ordinate, versus the channel number as the abscissa. Typical analyzers can analyze and store pulses simultaneously in 200, 256, 400, or 512 channels. If the counting rate is reasonably high, the lights flashing on the oscilloscope, due to the finite decay time of each light flash, will form a multi-point, gradually climbing, spectrum visible to the eye. At the conclusion of the counting, the spectrum is more readily visible as a static display.

When counting of a sample is completed, the stored data may be printed out in digital form (as the actual number of counts stored in each channel) and/or plotted out precisely on an x-y plotte. Calculations are then made by comparing the photopeak heights (or areas), above the Compton continuum base, of the peaks of interest with those of the standard samples. Usually, the $\gamma$ ray energy (known by spectrometer calibration with reference sources of known $\gamma$ energy, such as the 0.662-Mev $\gamma$ of $Cs^{137}$ — thus providing a conversion factor to convert channel number into $\gamma$ energy) suffices to identify a given radioisotope. If necessary, the half life may also be determined, as an additional confirmation of identity, by counting the sample at two or more decay times and then calculating the half life of the component in question from the decline of the photopeak counting rate with decay time.

Again, due to requirements of brevity, details concerning live-time circuits, deam-time meters, counting-rate gain shift, memory splitting, spectrum stripping, and computer calculations will have to be omitted from this presentation. A short mention of costs is, however, in order. A typical multichannel PHA costs about 10,000 dollars. However, the various desirable readout attachments (printer, plotter, spectrum stripper) usually add another 4,000 to 8,000 dollars to the cost.

## IV. Neutron sources

Two principal sources of neutrons, capable of producing neutron fluxes high enough to be of practical use in activation analysis work, are in widespread use today. These are (1) lower-cost, modest flux, small accelerators and (2) more expensive, very high flux research reactors. The main features of these two types of neutron sources are discussed briefly below.

### a) Accelerators

The accelerators particularly useful in this type of work are modest energy (0.1- to 2-Mev) deuteron accelerators, costing in the range of about 10,000 to 40,000 dollars. Generally the lower cost ones are small Cockcroft-Walton machines, operating at potentials of 100 to 150 kilovolts, costing 10,000 to 20,000 dollars, and producing from $10^{10}$ to $10^{11}$ 14-Mev neutrons per second (total output) from a fresh tritium target, via the $H^3(d,n)He^4$ reaction. The 14-Mev neutron flux obtainable at the location of a sample (perhaps one inch from the water-cooled tritium target) is then in the range of $10^8$ to $10^9$ neutrons/cm.$^2$-sec. Such high-energy neutrons are of value for producing $(n,p)$, $(n,a)$, and $(n,2n)$ reactions in many nuclei, as opposed to the $(n,\gamma)$ reactions produced by thermal neutrons. Certain elements, such as oxygen and nitrogen, can esentially only be activated by fast neutrons, e. g., oxygen by the $O^{16}(n,p)N^{16}$ reaction and nitrogen by the $N^{14}(n,2n)N^{13}$ reaction. Also, certain other elements, such as F, Si, P, S, Cr, and Fe, are activated more readily by fast neutrons than by thermal neutrons. A photograph of such an accelerator is shown in Fig. 7.

Fig. 7. Photograph of a small Cockcroft-Walton deuteron accelerator used to produce 14-Mev neutrons

If such an accelerator is to be used also as a source of thermal neutrons, the target must be surrounded by a foot or two of suitable moderator — a material such as water or paraffin that efficiently slows down the 14-Mev neutrons by nuclear collisions. Samples are then placed in openings in the moderator, fairly close to the tritium target. Such accelerators produce thermal neutron fluxes more in the range of $10^7$ to $10^8$ neutrons/cm.²-sec.

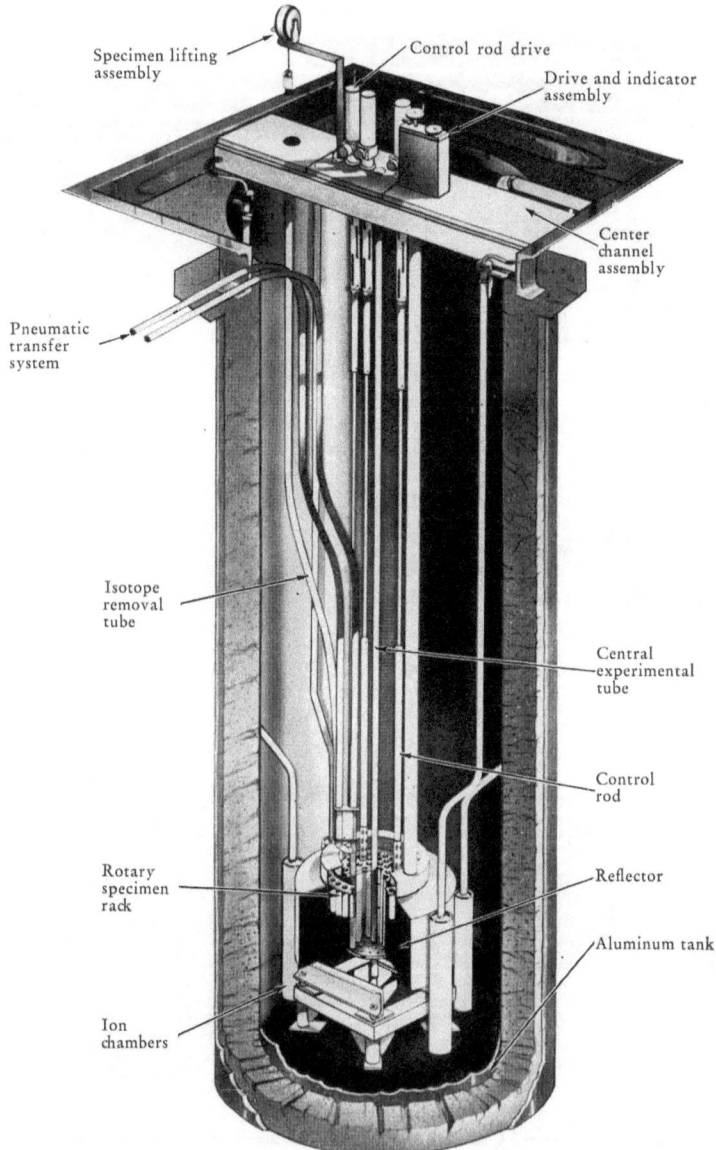

Specimen lifting assembly

Control rod drive

Drive and indicator assembly

Center channel assembly

Pneumatic transfer system

Isotope removal tube

Central experimental tube

Control rod

Rotary specimen rack

Reflector

Aluminum tank

Ion chambers

Fig. 8. Cutaway diagram of TRIGA Mark I reactor

One serious limitation of neutron sources which utilize the H³$(d,n)$He⁴ reaction is the limited target lifetime. Such targets (typically containing one to five curies of tritium per square inch of target) decline rather rapidly in their neutron output with continued use. At full-power operation they typically decrease in output by about a factor of 2 per hour of use. Most of the loss in output appears to be due to loss of tritium from the target.

High-energy neutrons are very penetrating and can be dangerous to personnel. Hence, a unit that produces $10^{11}$ 14-Mev neutrons per second needs to have something of the order of five feet of concrete shielding between the target and personnel.

A somewhat more expensive ($\sim$ 40,000 dollars) accelerator of the Van de Graaff type is very useful for the production of somewhat higher ($10^9$ neutrons/cm.$^2$-sec.) fluxes of thermal neutrons, and is free of target lifetime problem. This machine bombards a water-cooled beryllium target with 2-Mev deuterons, producing low-energy neutrons via the $Be^9(d,n)B^{10}$ reaction.

### b) Nuclear reactors

The nuclear reactor is by far the most prolific source of neutrons. Most reactors produce neutrons by means of the slow neutron fission of uranium-235. The most common type of modern research reactor is the "swimming pool" type. These generally consist of a cluster of uranium fuel elements (perhaps 50 to 100) located in a pool of water perhaps 20 feet deep. The cluster is somewhat more than a critical mass, and is shut down (made subcritical) by the deep insertion into the core of several boron or cadmium control rods. On startup, these are withdrawn sufficiently to make the core slightly supercritical. When the power level, and hence neutron flux, have climbed to the desired level, the reactor is steadied out at that level by slight insertion of the control rods to keep the core just critical. The critical mass of such a reactor is usually of the order of two kilograms of $U^{235}$. This is present in the fuel rods as somewhat $U^{235}$-enriched uranium, alloyed with some element such as zirconium, and clad in aluminum or stainless steel. One type of research reactor, the TRIGA (used by the authors) employs a partially hydrided U-Zr type of fuel element. The hydrided fuel element imparts the desired feature of intrinsic safety, and also enables one, where desired, to pulse the reactor to levels of the order of a billion watts (giving a short pulse with fluxes somewhat greater than $10^{16}$ neutrons/cm.$^2$-sec.). A cutaway diagram of the TRIGA reactor is shown in Fig. 8, and a photograph of the core, looking down through the water pool, is shown in Fig. 9.

Commercially available research reactors typically operate steadily in the 10 to 1,000 kilowatt power range, producing thermal neutron fluxes, respectively, in the range of $10^{11}$ to $10^{13}$ neutrons/cm.$^2$-sec., and costing, typically, in the range of 150,000 to 300,000 dollars. The 250-kilowatt TRIGA Mark I reactor used by the authors in activation analysis studies, for example, produces thermal neutron fluxes of $1.8 \times 10^{12}$ neutrons/cm.$^2$-sec. in the 40-tube rotating specimen rack just outside the core, inside of the graphite reflector, and of $3.5 \times 10^{12}$ neutrons/cm.$^2$-sec. in the outer fuel-element ring, where the pneumatic tube terminates. The latter is used where very short-lived induced activities (ones with half lives of the order of seconds or minutes) are of interest. With it, samples can be shot into a vacant fuel element position in 3 seconds, then ejected into the counting room, when desired, in 3 seconds. Such samples are thus activated one at a time, but many samples can be activated and counted in rapid succession.

Fig. 9. Photograph of TRIGA Mark I reactor, looking down at core through the water pool

Where the induced activities of interest have longer half lives (hours, days, weeks), it is usually desirable to activate them for longer periods of time — in order to increase the activity, as can be seen from Fig. 1. Here, however, many samples can be activated simultaneously. The TRIGA rotating specimen rack has 40 tubes. If one sample is placed in each tube, and the rack is rotated around the core during an irradiation, each sample receives exactly the same average dose of neutrons. Each tube can hold as

Table I. *Estimated neutron-activation sensitivities for irradiations of one hour or less in a thermal-neutron flux of $1.8 \times 10^{12}$*

| Element | Radionuclide measured | Radiochemical beta-ray sensitivity μg. | Instrumental gamma-ray sensitivity μg. | Element | Radionuclide measured | Radiochemical beta-ray sensitivity μg. | Instrumental gamma-ray sensitivity μg. |
|---|---|---|---|---|---|---|---|
| Ag | 2.3-m $Ag^{108}$ | 0.005 | 0.005 | Na | 15-h $Na^{24}$ | 0.005 | 0.005 |
|  | 24-s $Ag^{110}$ | — | 0.0001 | Nb | 6.6-m $Nb^{94m}$ | 0.005 | 1 |
| Al | 2.3-m $Al^{28}$ | 0.1 | 0.01 | Nd | 11.6-d $Nd^{147}$ | 0.1 | 0.1 |
| As | 27-h $As^{76}$ | 0.001 | 0.005 | Ni | 2.6-h $Ni^{65}$ | 0.05 | 0.5 |
| Au | 2.7-d $Au^{198}$ | 0.0005 | 0.0005 | Os | 31-h $Os^{193}$ | 0.05 | — |
| Ba | 85-m $Ba^{139}$ | 0.05 | 0.1 | P | 14.5-d $P^{32}$ | 0.5 | — |
| Bi | 5.0-d $Bi^{210}$ | 0.5 | — | Pb | 3.3-h $Pb^{209}$ | 10 | — |
| Br | 17.6-m $Br^{80}$ | 0.005 | 0.005 | Pd | 4.8-m $Pd^{109m}$ | — | 0.05 |
|  | 36-h $Br^{82}$ | 0.005 | 0.01 |  | 13.6-h $Pd^{109}$ | 0.0005 | 5 |
| Ca | 8.8-m $Ca^{49}$ | 1.0 | 5 | Pr | 19-h $Pr^{142}$ | 0.0005 | 0.05 |
| Cd | 54-h $Cd^{115}$ | 0.05 | 0.5 | Pt | 31-m $Pt^{199}$ | 0.05 | 0.1 |
| Ce | 32-d $Ce^{141}$ | 1 | 1 |  | 3.2-d $Au^{199}$ | 0.1 | 0.1 |
|  | 32-h $Ce^{143}$ | 0.1 | 0.1 | Rb | 18.6-d $Rb^{86}$ | 0.05 | 5 |
| Cl | 37-m $Cl^{38}$ | 0.01 | 0.1 | Re | 91-h $Re^{186}$ | 0.001 | 0.05 |
| Co | 10.3-m $Co^{60m}$ | 0.005 | 0.1 |  | 17-h $Re^{188}$ | 0.0005 | 0.001 |
|  | 5.3-y $Co^{60}$ | 0.5 | 0.5 | Rh | 4.4-m $Rh^{104m}$ | 0.001 | 0.0005 |
| Cr | 27-d $Cr^{51}$ | (no. $\beta$) | 1 |  | 4.2-s $Rh^{104}$ | — | 0.01 |
| Cs | 2.2-y $Cs^{134}$ | 0.5 | 0.5 | Ru | 40-d $Ru^{103}$ | 0.5 | 1 |
| Cu | 12.8-h $Cu^{64}$ | 0.001 | 0.001 |  | 4.5-h $Ru^{105}$ | 0.01 | 0.05 |
|  | 5.1-m $Cu^{66}$ | 0.01 | 0.05 | S | 87-d $S^{35}$ | 10 | — |
| Dy | 2.3-h $Dy^{165}$ | 0.000001 | 0.000005 |  | 5.0-m $S^{37}$ | 5 | 200 |
| Er | 9.4-d $Er^{169}$ | 0.1 | — | Sb | 2.8-d $Sb^{122}$ | 0.005 | 0.01 |
|  | 7.5-h $Er^{171}$ | 0.001 | 0.001 | Sc | 84-d $Sc^{46}$ | 0.01 | 0.05 |
| Eu | 9.3-h $Eu^{152m}$ | 0.000005 | 0.0005 | Se | 120-d $Se^{75}$ | (no. $\beta$) | 5 |
| F | 11-s $F^{20}$ | — | 1 | Si | 2.6-h $Si^{31}$ | 0.05 | 500 |
| Fe | 45-d $Fe^{59}$ | 50 | 200 | Sm | 47-h $Sm^{153}$ | 0.0005 | 0.005 |
| Ga | 14-h $Ga^{72}$ | 0.005 | 0.005 | Sn | 9.5-m $Sn^{125m}$ | 0.5 | 0.5 |
| Gd | 18-h $Gd^{159}$ | 0.01 | 0.05 | Sr | 2.8-h $Sr^{87m}$ | 0.005 | 0.005 |
| Ge | 82-m $Ge^{75}$ | 0.005 | 0.05 |  | 64-d $Sr^{85}$ | 50 | 50 |
|  | 11-h $Ge^{77}$ | 0.5 | — | Ta | 112-d $Ta^{182}$ | 0.05 | 0.5 |
| Hf | 19-s $Hf^{179m}$ | — | 1 | Tb | 72-d $Tb^{160}$ | 0.05 | 0.1 |
| Hg | 65-h $Hg^{197}$ | (no. $\beta$) | 0.01 | Te | 25-m $Te^{131}$ | 0.05 | 0.05 |
| Ho | 27-h $Ho^{166}$ | 0.0001 | 0.0001 |  | 8-d $I^{131}$ | 5 | 1 |
| I | 25-m $I^{128}$ | 0.005 | 0.01 | Th | 27-d $Pa^{233}$ | 0.05 | 0.05 |
| In | 54-h $In^{116m}$ | 0.00005 | 0.0001 | Ti | 5.8-m $Ti^{51}$ | 0.5 | 0.05 |
| Ir | 19-h $Ir^{194}$ | 0.0001 | 0.001 | Tm | 127-d $Tm^{170}$ | 0.01 | 0.1 |
| K | 12.5-h $K^{42}$ | 0.05 | 0.5 | U | 2.3-d $Np^{239}$ | 0.005 | 0.005 |
| La | 40-h $La^{140}$ | 0.001 | 0.005 | V | 3.8-m $V^{52}$ | 0.005 | 0.001 |
| Lu | 3.7-h $Lu^{176}$ | 0.00005 | 0.00005 | W | 24-h $W^{187}$ | 0.001 | 0.005 |
|  | 6.8-d $Lu^{177}$ | 0.0005 | 0.005 | Yb | 4.2-d $Yb^{175}$ | 0.001 | 0.01 |
| Mg | 9.5-m $Mg^{27}$ | 0.5 | 0.5 | Zn | 14-h $Zn^{69m}$ | 0.1 | 0.1 |
| Mn | 2.6-h $Mn^{56}$ | 0.00005 | 0.00005 | Zr | 17-h $Zr^{97}$ | 1 | 1 |
| Mo | 15-m $Mo^{101}$ | 0.1 | 5 |  |  |  |  |
|  | 67-h $Mo^{99}$ | 0.5 | 0.1 |  |  |  |  |

many as six small sample vials, thus even 240 samples can be activated at the same time. With the longer half lives, there is then ample time to count them leisurely, one at a time.

## V. Detection sensitivities attainable

Where high thermal-neutron fluxes are available, as in research reactors ($10^{11}$ to $10^{13}$ neutrons/cm.²-sec.), neutron-activation analysis becomes the most sensitive known method for the quantitative detection of low concentrations of more than half of the elements in the periodic system. For example, in Table I, taken from BUCHANANS' work (1961) with the 250-kw TRIGA Mark I reactor at General Atomic, detection limits for 69 elements are listed, all for a thermal neutron flux of $1.8 \times 10^{12}$ neutrons/cm.²-sec. and a maximum irradiation time of one hour. The most sensitive element listed is dysprosium ($10^{-6}$ μg. detectable), the least sensitive one listed is iron (50 μg. detectable), and the median sensitivity is 0.01 μg. Thus, with one-gram samples, a median sensitivity of detection is 0.01 p.p.m., or 10 p.p.b. Since even 10-gram samples can be utilized, a median limit of 1 p.p.b. is then attainable.

The limits shown in Table I are calculated ones, based on very reasonable assumptions on counting. Most have been, and all can be, attained in practice, in the absence of appreciable interferences. Even where interferences limit the ultimate sensitivity attainable purely instrumentally (by nondestructive analysis, employing gamma-ray spectrometry), the interferences can be eliminated, and the calculated sensitivities achieved, if one resorts to radiochemical separations with carriers after the activation.

Whereas close to the detection limits the precision falls off rapidly, the method is quite precise and accurate at levels appreciably above the limits of detection — soon reaching precisions and absolute accuracies in the range of 1 to 5 percent of the value.

Since the activity generated in a particular element is directly proportional to the neutron flux employed (other factors being equal), it is evident that the sensitivities attainable with lower neutron fluxes are not as good as those shown in Table I. Thus, at a thermal neutron flux of $1.8 \times 10^8$ neutrons/cm.²-sec., representative of that attainable with one of the small accelerators described earlier, the sensitivity limits of Table I must all be multiplied by $10^4$. Thus, with the lower flux of a small accelerator, the median sensitivity will be about 100 μg., instead of 0.01 μg., and the median concentration sensitivity in a one-gram sample will be 100 p.p.m., instead of 0.01 p.pm. The small accelerator can still be very useful for many elements, because of the speed of the instrumental method, but is no longer a method of extraordinary sensitivity. The purely instrumental method, employing such accelerators, has been described extensively by GUINN and WAGNER (1960) and by GUINN (1961 a, b, and c).

## VI. Experimental procedure

In general, the experimental procedure of instrumental neutron-activation analysis, to which all of the remaining discussion in this paper will be limited, is quite simple. Samples to be analyzed are merely placed in weighed polyethylene vials, tightly capped or heat-sealed, weighed, and then irradiated one-at-a-time (for short-lived activities) or in larger numbers (for longer-lived activities). Suitable sample weights range from milligrams to perhaps 10 grams, depending upon the amount of sample available, the neutron flux available, irradiation time considered reasonable, and sensitivity needed. In any given series of samples, all should have about the same volume, the same as the volume of standard solution of each element of interest that is simultaneously irradiated. The constancy of sample volumes and standard solution volumes is necessary so that all will be exposed to the same neutron flux, and will be counted at the same geometry. The bulk densities of the samples should be somewhere near that of the standard solution (normally aqueous solutions and hence of unit density) so that self-absorption and self-scattering of gamma rays emitted by the samples will be small and comparable with that in the standard solutions. Fortunately, the bulk density limitation is not a very severe one, bulk densities in the range of 0.5 to 1.5 g./ml. being acceptable. Agricultural, biological, and medical samples usually have bulk densities in this range.

Samples and standard solutions are activated at exactly the same flux, for exactly the same time, and are then counted under exactly the same conditions. Thus the photopeak counting rate of a given induced activity of interest merely needs to be corrected (via equation 5) to the same decay time as that at which the standard solution of that element was counted (or both corrected to $t_o$), and compared with that of the standard. Then,

$$\frac{\text{photopeak cpm. (sample)}}{\text{photopeak cpm. (standard)}} = \frac{\mu\text{g. element in sample}}{\mu\text{g. element in standard}} \qquad (10)$$

In this equation, *cpm* is the abbreviation for counts per minute. Three of the four quantities in equation 10 are then known, and hence the fourth — the desired value, $\mu$g. of element in sample, is readily calculated. This comparator method, involving activation and counting of a standard sample of each element of interest, is virtually always used. It eliminates the necessity of knowing the values of several of the terms in equation 3, namely, of $f$ and $\sigma$. This is fortunate, since values of flux and reaction cross section are often only known approximately. Whatever their true values may be, however, they are the same for sample and standard. The comparator method also eliminates the necessity of knowing the exact counting efficiency, which would otherwise have to be known in order to convert observed photopeak counts per minute to the total isotope disintegrations per second of equation 3. As long as sample and standard are counted identically, their photopeak cpm values are in exactly the same ratio as their dps values.

Although the polyethylene vials are fairly pure hydrocarbon, and neither C nor H becomes detectably activated, they do contain traces of certain impurity elements (such as Na, Cl, Mn, Al, Ti). For measurements at very low levels it is thus desirable to either correct the sample observed spectrum for the contributions from the vial itself (by activating and counting an empty vial), or else transferring the activated sample to a fresh vial before counting. The sample and standard vials should also be flushed with a suitable gas, such as helium, nitrogen, or $CO_2$, before sealing and activating. If this is not done, the 0.9 percent argon present in the air in the vial will become appreciably active, forming 1.8-hour $A^{41}$. This radioactive argon emits a 1.29-Mev gamma ray, thus unnecessarily complicating the observed spectra. Alternatively, the samples (and standard) vials can be opened and swept out with air after activation, before counting, to remove the $A^{41}$ present.

A stock of standard solutions of the various elements of interest, of suitable and accurately known concentrations ($\mu$g./ml. or mg./ml.), is prepared, kept on hand, and used as needed. Compounds that are water-soluble and in which only the element of interest becomes appreciably activated by thermal neutrons are used in their preparation. Metal nitrates, for example, are often used for metal ion standards (Al, Mn, Fe, Cu, etc.), as well as the ammonium salts of the halogens for halogen standards (F, Cl, Br, I). Many of these are not suitable as primary standards, and hence the solutions prepared must be then standardized by regular chemical analytical methods. If the desired level is too low for ready standardization by chemical means, a more concentrated solution is first prepared and standardized, then carefully diluted to the desired strength with distilled, deionized water.

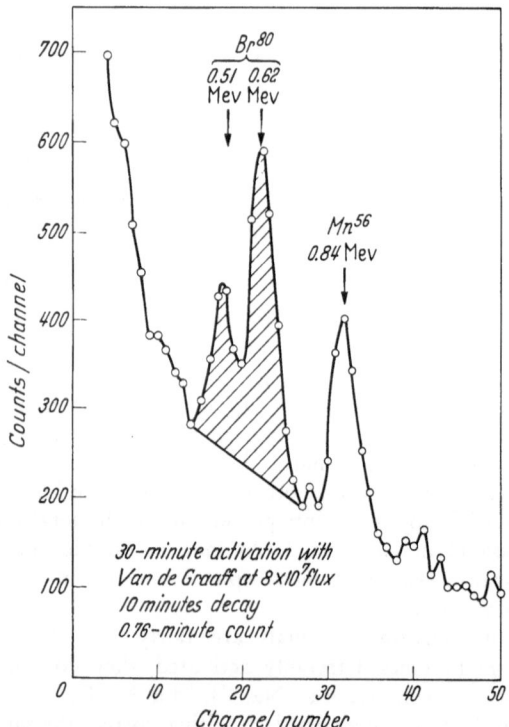

Fig. 10. Gamma-ray spectrum of neutron-activated sample of nematocide-treated bush beans

## VII. Applications in the determination of pesticide residues

Agricultural samples readily lend themselves to analysis by instrumental neutron-activation analysis. Their bulk densities are suitable for direct comparison with aqueous standard solutions. They do not contain appreciable amounts of high neutron absorption cross-section elements, such as lithium, boron, cadmium, or certain rare earth elements, so neutron self-shielding is not a problem. Usually, plenty of sample is available, compared with the amount desired ($\sim$ one gram). Often large numbers of samples need to be analyzed even in a single study, and hence the high speed of the method is attractive. Very good detection sensitivities can be achieved in the reactor for a number of the elements of particular interest in pesticide residue studies, for example, Cl and Br (and also for numerous elements of interest in the study of trace element roles in plant metabolism).

### a) Bromine residues in crops from nematocides

The first application of the instrumental neutron-activation analysis technique to pesticide residue problems was that carried out by GUINN and POTTER (1962). They determined total Br in a variety of crops, employing at first a low flux ($10^8$ neutrons/cm.²-sec.) Van de Graaff accelerator neutron source and gamma-ray spectrometer detection of 18-minute Br[80]. The bromine-80 produced in 20-gram samples, in a 30-minute irradiation at this flux, enabled them to detect Br levels down to about 10 p.p.m. Bromine-80 was detected in the gamma-ray spectra from its 0.62-Mev gamma-ray photopeak. An illustrative spectrum from their study is shown in Fig. 10. In the second part of their work, they employed a TRIGA reactor as the neutron source (at a flux of about $10^{12}$ neutrons/cm.²-sec.), and 30-minute irradiations of one-gram samples. After about two days decay, the gamma-ray spectra of the samples were measured, and the Br contents determined from the 0.77-Mev gamma-ray photopeak of 35.9-hour Br[82]. Even in the presence of interferences from other radioisotopes present, Br levels as low as one p.p.m. could be determined.

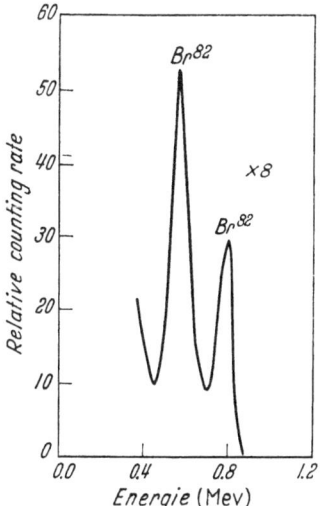

In this work, both untreated plants, and plants grown in soil treated with various amounts of the nematocide, 1,2-dibromo-3-chloropropane, were analyzed for Br. The natural levels of Br, in untreated plants were usually of the order of one p.p.m. In the soil-fumigated samples much higher levels (often 10 to 100 p.p.m.) were encountered. As can be seen from Table I, the sensitivities for detection of Br by gamma ray spectrometry, at the reactor flux, are 0.005 $\mu$g. (Br[80]) and 0.01 $\mu$g. (Br[82]).

Fig. 11. Gamma-ray spectrum of neutron-activated sample of raw orange juice

In ensuing work employing this technique, CASTRO and SCHMITT (1962) analyzed many citrus crop samples for nematocide Br residues, using the TRIGA reactor and Br[82] detection. The gamma-ray spectrum of one of their samples, orange juice containing about 5 p.p.m. of Br, is shown in Fig. 11. They also determined Cl, Mn, Na, and K in such samples.

LINDGREN, GUNTHER, and VINCENT (1962) and LINDGREN (1962) employed the same technique for determining Br residues of a different type — those produced in wheat by methyl bromide fumigation during storage. The $Br^{82}$ analyses were carried out for them with a TRIGA reactor by the activation-analysis group at General Atomic.

In the authors' laboratory the method has since been used to determine Br residues from nematocides, and from methyl bromide fumigations, in hundreds of samples.

### b) Chlorinated pesticide residues in foods

Instrumental neutron-activation analysis was first applied to this problem by SCHMITT and ZWEIG (1962). They activated organic extracts of butterfat and thus determined the total organic chloride content of the butterfat. They employed a TRIGA reactor and determined the Cl from the 1.59-Mev and 2.16-Mev gamma-ray photopeaks of 37.3-minute $Cl^{38}$. From Table I the normal limit of detection of Cl by this means is seen to be about 0.1 $\mu$g. This method has subsequently been used by the General Atomic activation analysis group for organic chloride determination on hundreds of agricultural samples. Some of this work is described by BUCHANAN and GUINN (1963).

### c) Mercury in wheat

An interesting current problem is that of determining traces of mercury in wheat. The problem arises from the fact that some distributors and farmers apparently dump left-over seed wheat, which contains about 10 p.p.m. of Hg (to prevent fungus attack during germination in the soil), in new crop wheat. This is in violation of FDA rulings, but is very difficult to detect, since, typically, batches of the resulting contaminated wheat may only contain perhaps 0.01 to 1 p.p.m. levels of mercury. In preliminary studies by the General Atomic activation-analysis group, it has been shown that mercury can be determined by instrumental neutron-activation analysis, with the reactor, down to levels of about 0.1 p.p.m. — employing detection of the 0.077-Mev gamma ray of 65-hour $Hg^{197}$. With radiochemical separations, sensitivities down to 0.001 p.p.m. Hg are attainable.

### Acknowledgements

The authors wish to express their appreciation for the contributions and helpful discussions provided by their colleagues in the General Atomic laboratory, especially R. R. Ruch, Miss Diane M. Fleishman, J. C. Migliore, J. E. Lasch, D. A. Olehy, and J. D. Buchanan; and to their collaborators in other laboratories, namely, C. E. Castro, F. A. Gunther, and D. L. Lindgren (University of California, Riverside), G. Zweig (University of California, Davis), and J. C. Potter (Shell Development Company, Modesto).

### Summary

Neutron-activation analysis is a method of elemental analysis in which various elements present in a sample are detected quantitatively by means of the radiations emitted by neutron-induced radioactivities in the sample.

Each element can form one or more radioisotopes upon capture or other types of interaction with neutrons. Each radioisotope can be identified by means of the type and energy of its radiations, and the half life with which it decays.

The basic mathematical equation of activation analysis is discussed in some detail. Also, the experimental procedures of sample preparation, neutron activation, sample treatment after activation, gamma-ray spectrometry, and calculational methods are discussed.

Tables of ultimate (interference-free) neutron-activation sensitivities of detection, for a typical research-reactor neutron flux, are given. Specific applications of neutron-activation analysis in the pesticide-residue field are discussed. As examples, the results of various investigators for the determination of trace amounts of bromine in a wide variety of crops and foodstuffs, pesticide-treated and untreated, have been discussed. The concentrations of total organic chloride in milk butterfat and forage samples have been determined simply and accurately by neutron activation analysis.

The various useful types of neutron sources and radiation detection instruments are described in some detail. A number of figures, showing a nuclear reactor, neutron generator, gamma ray spectrometer, and some representative gamma-ray spectra, are included.

The principal advantages of neutron-activation analysis, as compared with other methods of determining trace element concentrations, are:

(1) Extreme sensitivity — sensitivities ranging from fractions of a part per billion up to parts per million are attainable for most elements. In many cases these sensitivities are orders of magnitude better than those achieved by spectrographic or colorimetric methods.

(2) The problem of sample contamination by impurities in chemical reagents used in the analyses is eliminated, since normally no reagents are added until after sample activation, if at all.

(3) Often samples can be analyzed nondestructively, by simply irradiating with neutrons and then examining the activated sample by means of gamma-ray spectrometry. Suitable choice of irradiation time and decay time prior to counting simplify the observed gamma-ray spectrum and emphasize the desired activity. Such analyses are very rapid.

### Résumé *

L'analyse par activation par les neutrons est une méthode d'analyse élémentaire dans laquelle plusieurs éléments d'un échantillon sont détectés quantitativement au moyen des radiations émises par les radioactivités induites par les neutrons dans l'échantillon. Chaque élément peut former un ou plusieurs radioisotopes par capture ou grâce à d'autres types d'interaction avec les neutrons. Chaque radioisotope peut être identifié au moyen du type et de l'énergie de ses radiations et de la période avec laquelle il décroît.

---

* Traduit par R. Mestres.

L'équation mathématique fondamentale de l'analyse par activation est assez largement discutée ainsi que les méthodes expérimentales de préparation de l'échantillon, de l'activation, du traitement de l'échantillon après activation, de la spectrométrie gamma et les méthodes de calculs.

Les tables des sensibilités limites (exemptes d'interférences) de la détection par activation par les neutrons sont données pour le flux de neutrons d'un réacteur type de recherche. Les applications spécifiques de l'analyse par activation par les neutrons dans le domaine des résidus de pesticides sont commentées. A titre d'exemples sont donnés les résultats des travaux de plusieurs chercheurs sur le dosage des traces de Brome dans une grande variété de récoltes et de matières alimentaires. Les concentrations en chlore organique total dans le beurre et dans des fourrages ont été déterminées simplement et avec précision grâce à l'analyse par activation aux neutrons.

Les divers types de sources de neutrons et d'appareils détecteurs de radiations sont décrits avec quelques détails. Des figures montrant un réacteur nucléaire, un générateur de neutrons, un spectromètre gamma et quelques spectres représentatifs de rayons gamma illustrent ce texte.

Les principaux avantages de l'analyse par activation aux neutrons sur les autres méthodes d'analyses de traces sont: — 1. la sensibilité extrême: des sensibilités allant de quelques fractions de partie par milliard jusqu'aux parties par million sont possibles pour la plupart des éléments. Dans beaucoup de cas, ces sensibilités sont d'un ordre de grandeur supérieur à celui donné par les méthodes spectrographiques ou colorimétriques. — 2. Le problème de la contamination de l'échantillon par des impuretés introduites par les réactifs chimiques est éliminé puisque l'on n'ajoute normalement aucun réactif avant l'activation de l'échantillon.

— 3. Les échantillons peuvent être souvent analysés sans être détruits, par simple irradiation avec des neutrons et examen de l'échantillon activé par spectrométrie gamma. Des durées convenables d'irradiation et un comptage après une durée de décroissance judicieusement choisie simplifient le spectre gamma obtenu et exaltent l'activité désirée. De telles analyses sont très rapides.

## Zusammenfassung *

Neutronenaktivierungs-Analyse ist eine Methode der Elementaranalyse, bei der verschiedene in einer Analysenprobe vorliegende Elemente quantitativ mittels der Strahlung bestimmt werden, die von einer neutroneninduzierten Radioaktivität der Probe emittiert wird. Jedes Element kann ein Radioisotop oder mehrere solcher Isotope bilden, entweder durch Einfangen von Neutronen oder durch sonstige Wechselbeziehungen mit Neutronen. Jedes Radioisotop kann durch Art und Energie der von ihm ausgesandten Strahlung und auf Grund seiner Halbwertzeit identifiziert werden.

Die grundlegende mathematische Gleichung der Aktivierungsanalyse wird in einigen Einzelheiten diskutiert, ebenfalls werden diskutiert die

---

* Übersetzt von H. FREHSE.

praktische Durchführung der Probenaufbereitung, die Neutronenaktivierung, Probenbehandlung nach der Aktivierung, Gammastrahlenspektrometrie und Berechnungsverfahren.

Tabellen für die letztlich erzielbaren, störfreien Neutronenaktivierungs-Nachweisempfindlichkeiten für den Neutronenstrom eines typischen Forschungsinstruments werden angegeben: spezifische Anwendungsformen der Neutronenaktivierungs-Analyse auf dem Gebiet der Pestizidrückstände werden besprochen. Als Beispiel sind die Ergebnisse verschiedener Bearbeiter bei der Bestimmung von Spurenmengen von Brom in einer großen Zahl von Pflanzen und Lebensmitteln, die sowohl mit Pestizid behandelt als auch nicht behandelt waren, angegeben. Die Mengen des gesamten organisch gebundenen Chlors in Milchfett und Futterproben ließen sich einfach und exakt durch Neutronenaktivierungs-Analyse bestimmen.

Die verschiedenen brauchbaren Typen von Neutronenquellen und Strahlungsdetektoren werden in einigen Einzelheiten diskutiert; gleichzeitig werden Abbildungen eines Kernreaktors, Neutronengenerators, Gammastrahlenspektrometers und einiger typischer Gammastrahlenspektren wiedergegeben.

Die hauptsächlichen Vorteile der Neutronenaktivierungs-Analyse beim Vergleich mit anderen Methoden der Spurenanalyse sind:

1. extreme Empfindlichkeit; für die meisten Elemente lassen sich Nachweise im Bereich von ppb-Bruchteilen bis zu mehreren ppm führen. In vielen Fällen ist die Empfindlichkeit um mehrere Zehnerpotenzen besser als die spektrographisch oder kolorimetrisch erzielbare.

2. Das Problem der mit den Reagenzien eingeschleppten Verunreinigungen ist ausgeschaltet, da normalerweise eine Reagenzienzugabe entweder erst nach der Aktivierung der Probe oder gar nicht erfolgt.

3. Oft können Proben ohne Zerstörung analysiert werden, einfach durch Bestrahlen mit Neutronen und Untersuchung der aktivierten Probe mittels Gammastrahlenspektrometrie. Geeignete Auswahl von Bestrahlungs- und Zerfallszeiten vor der Zählung vereinfacht das beobachtete Gammastrahlenspektrum und unterstreicht die gewünschte Aktivität. Solche Analysen arbeiten sehr schnell.

## References

BUCHANAN, J. D.: Activation analysis with a TRIGA Reactor. Proc. 1961 Int'l. Conf. on Modern Trends in Activation Analysis, College Station, Texas, Dec. 15—16, p 72 (1961).

—, and V. P. GUINN: Analysis of foods by neutron activation techniques. Food Technol. 17, 17 (1963).

GUINN, V. P., and C. D. WAGNER: Instrumental neutron activation analysis. Anal. Chem. 32, 317 (1960).

GUINN, V. P.: Neutron activation analysis. Int'l. Sci. and Technol., Prototype Issue, 74 (1961 a).

— Instrumental neutron activation for rapid, economical analysis. Nucleonics 19, 81 (1961 b).

— New developments in instrumental activation analysis — accelerators and analyzers. Proc. 1961 Int'l. Conf. on Modern Trends in Activation Analysis, College Station, Texas, Dec. 15—16, p. 126 (1961 c).

Guinn, V. P., and J. C. Potter: Determination of total bromine residues in agricultural crops by instrumental neutron activation analysis. J. Agr. Food Chem. 10, 232 (1962).

Castro, C. E., and R. A. Schmitt: Direct elemental analysis of citrus crops by instrumental neutron activation. A rapid method for total bromide, chloride, manganese, sodium, and potassium residues. J. Agr. Food Chem. 10, 236 (1962).

Lindgren, D. L., F. A. Gunther, and L. E. Vincent: Bromine residues in wheat and milled wheat fractions fumigated with methyl bromide. J. Econ. Entomol. 55, 773 (1962).

— Detection of Br residues in wheat, resulting from CH₃Br fumigations. Western Feed and Seed. June (1962).

Schmitt, R. A., and G. Zweig: Total organic chloride content in milk butterfat by a rapid method of neutron activation analysis. J. Agr. Food Chem. 10, 481 (1962).

# Subject index